Essential Law for Landowners a
Fourth Edition

Essential Law for Landowners and Farmers

fourth edition

Angela Sydenham
Birketts Solicitors, Ipswich

Bruce Monnington
Country Land and Business Association

Andrew Pym
Chartered Surveyor

Blackwell
Science

© 2002 by Blackwell Science Ltd,
a Blackwell Publishing Company
Editorial Offices:
Osney Mead, Oxford OX2 0EL, UK
 Tel: +44 (0) 1865 206206
Blackwell Science, Inc., 350 Main Street,
Malden, MA 02148-5018, USA
 Tel: +1 781 388 8250
Iowa State Press, a Blackwell Publishing
Company, 2121 State Avenue, Ames, Iowa
50014-8300, USA
 Tel: +1 515 292 0140
Blackwell Science Asia Pty, 54 University
Street, Carlton, Victoria 3053, Australia
 Tel: +61 (0)3 9347 0300
Blackwell Wissenschafts Verlag,
Kurfürstendamm 57, 10707 Berlin,
Germany
 Tel: +49 (0)30 32 79 060

The right of the Author to be identified as
the Author of this Work has been asserted in
accordance with the Copyright, Designs and
Patents Act 1988.

All rights reserved. No part of this
publication may be reproduced, stored in a
retrieval system, or transmitted, in any form
or by any means, electronic, mechanical,
photocopying, recording or otherwise,
except as permitted by the UK Copyright,
Designs and Patents Act 1988, without the
prior permission of the publisher.

First Edition published by Granada
Publishing – Technical Books Division 1980
Second Edition published by Collins
Professional and Technical Books 1987
Third Edition published by BSP Professional
Books 1990
Fourth Edition published by Blackwell
Science 2002

Library of Congress
Cataloging-in-Publication Data
is available

ISBN 0-632-05796-3

A catalogue record for this title is available
from the British Library

Set in 10/12pt Sabon
by DP Photosetting, Aylesbury, Bucks
Printed and bound in Great Britain by
MPG Books Ltd, Bodmin, Cornwall

For further information on
Blackwell Science, visit our website:
www.blackwell-science.com

Cover illustration courtesy of Nigel Farthing, Birketts Solicitors

In Memoriam
Michael Gregory

Contents

Preface to the fourth edition   xi

1 Introduction to Land Law   1
   1.1 General principles   1
   1.2 The legal estates   2
   1.3 Legal interests   3
   1.4 Equitable interests   4
   1.5 Registration of title   6

2 Farm Business Tenancies   12
   2.1 Introduction   12
   2.2 Meaning of farm business tenancy   12
   2.3 Termination of a farm business tenancy   17
   2.4 Rent   19
   2.5 Fixtures and compensation   23
   2.6 Dispute resolution   27
   2.7 Notices   29

3 Agricultural Holdings Act Tenancies   31
   3.1 Introduction   31
   3.2 Definitions   31
   3.3 The tenancy agreement   32
   3.4 Fixed equipment   32
   3.5 Rent   34
   3.6 Security of tenure   36
   3.7 Compensation   39
   3.8 Succession tenancies   42
   3.9 Disputes procedure   45

4 Residential Protection of Farm Workers   47
   4.1 Introduction   47
   4.2 The Rent (Agriculture) Act 1976   47
   4.3 The Housing Act 1988   53
   4.4 The Rent Act 1977   56
   4.5 The Protection From Eviction Act 1977   57
   4.6 Shorthold tenancies   57

5 Business Tenancies on Farms and Estates   60
   5.1 Introduction   60
   5.2 Agricultural Holdings Act tenancies   60
   5.3 Farm business tenancies   60

	5.4	New tenants	61
	5.5	The legal nature of a lease or tenancy	61
	5.6	The Landlord and Tenant Act 1954 Part II	62
	5.7	Stamp duty	67
	5.8	Registration	68
6	Public Access		69
	6.1	Introduction	69
	6.2	Public rights of way	69
	6.3	Limitations on land use causd by public rights of way	72
	6.4	Access to open land and registered common land	80
	6.5	Towns and village greens	84
7	Planning		86
	7.1	Introduction	86
	7.2	The need for planning permission	86
	7.3	The General Permitted Development Order	88
	7.4	Obtaining planning permission	96
	7.5	Specific planning issues	100
	7.6	Listed buildings, conservation areas and other designated property	103
	7.7	Appeals	108
	7.8	Enforcement of planning control	111
	7.9	The development plan	114
	7.10	Government guidance	117
8	Compulsory Purchase and Compensation		118
	8.1	Introduction	118
	8.2	The legislative background	118
	8.3	The power to acquire land compulsorily	120
	8.4	The assessment of compensation	124
	8.5	The Critchel Down rules	135
9	Utilities – Rights and Wayleaves		137
	9.1	Introduction	137
	9.2	Wayleaves	137
	9.3	Works in the highway land and private tracks	138
	9.4	Electricity	140
	9.5	Telecoms	144
	9.6	Gas	146
	9.7	Water resources and drainage	149
	9.8	Water and sewerage	152
	9.9	All other pipelines	156

Contents ix

10	Liability of the Occupier of Land	158
	10.1 Impact of land use	158
	10.2 Insurance	158
	10.3 Negligence	159
	10.4 Occupiers' liability	159
	10.5 Nuisance	162
	10.6 Strict liability – *Rylands* v. *Fletcher*	168
	10.7 Straying animals	169
	10.8 Trespass	171
11	Water and Watercourses	174
	11.1 Responsibility for water management	174
	11.2 Land acquisition	175
	11.3 Riparian rights	175
	11.4 Water abstraction	176
	11.5 Impounding water	180
	11.6 Stocking fishing lakes	181
	11.7 Land drainage and flood defence	181
12	Sporting Matters	185
	12.1 Sporting rights	185
	12.2 Protection of wildlife	187
	12.3 Game laws	189
	12.4 Poaching of game	189
	12.5 Firearms	191
	12.6 Fishing	193
	12.7 Poaching of fish	199
	12.8 The Environment Agency	200
	12.9 Hunting	200

Appendix to Chapter 12 202

13	Environment	207
	13.1 Introduction	207
	13.2 Water pollution	207
	13.3 Other pollution control measures	209
	13.4 The codes of practice	210
	13.5 Contaminated land	211
	13.6 Fly tipping	216
	13.7 Waste disposal	218
	13.8 The Hedgerow Regulations 1997	223
14	Milk Quotas	224
	14.1 Introduction	224
	14.2 The milk quota system	224
	14.3 Special quota	225

	14.4	Transfer of quota	225
	14.5	Leasing of milk quota	230
	14.6	Milk quota and tenancies	230
15	Employment		235
	15.1	Introduction	235
	15.2	Employment Tribunals	235
	15.3	Employees or self-employed workers	236
	15.4	Engaging a worker	238
	15.5	Breaches of contract	242
	15.6	Rights under the employment legislation	242
	15.7	Discipline	244
	15.8	Dismissal	244
	15.9	Redundancy	247
	15.10	Change in ownership of business	251
	15.11	Health and safety at work	253
16	Rating and Council Tax		254
	16.1	Rating	254
	16.2	Council Tax	261

Further Reading	265
Table of Cases	266
Table of Statutes	271
Table of Statutory Instruments	287
Index	290

Preface to the fourth edition

It is eleven years since the third edition of this book. There have been many changes to the law during that time, not least the introduction of the farm business tenancy.

Whilst acknowledging the debt we owe to Michael Gregory and Margaret Parrish, the original authors of the book, we have taken the opportunity in this edition to revise the format and extend the content. Chapters 1, 2, 5 and 13 cover new topics and there has been a complete rewrite of Chapters 3, 4, 6, 7, 8 and 9. The other chapters have been substantially updated. The book states the law at 31 January 2001. However, new legislation and cases mean that the law is constantly changing.

Therefore, readers should not rely on this book alone when trying to resolve their legal problems. Moreover so short a book, covering such a wide range of topics, can be no more than an introduction to the areas of law with which the landowner and farmer is likely to be involved. It is important therefore that professional advice should be taken from lawyers and land agents on individual problems.

All three authors have worked for the Country Landowners and Business Association.

In our capacity as legal advisers and as a rural practice surveyor we have answered numerous questions relating to land ownership and use. As in previous editions we have concentrated on those problems which cause most trouble and generate most queries. While many issues affecting land are decided in the courts on the basis of the law, some are dedicated by authorities or inspectors exercising their discretion, in the light of government policy, having heard wide ranging evidence. For that reason the chapters on planning, compulsory purchase and utilities have been written by a surveyor.

Our hope is that this book will prove useful to landowners, farmers and their advisers, students studying agricultural law and their teachers.

Chapter 1

Introduction to Land Law

1.1 General principles

Land law is a complicated area of the law. There are several reasons for this. Land is permanent. Many of the rules were originally developed in feudal times. Although the law has changed to meet the needs of modern society, it reflects not a clean sweep but a continuous historical growth over many centuries.

Land being a scarce resource can never remain wholly within the control of private individuals. Economic and political factors have led to state intervention in many areas most notably in housing and planning law.

A further complication is the abstract theory of land law which enables many estates and interests, both present and future, to exist in the same piece of land.

When William I conquered England he considered himself to be the owner of all English soil. In order to reward his supporters he granted them rights over land in return for certain services. One baron might provide ten armed horsemen each year for the king, another might perform ceremonial rights. These tenants in chief (about 1,500 were recorded in 1086 at the time of the Domesday Book) could in turn grant rights to others for services rendered to them, and so on. There emerged a feudal pyramid with the king at the apex and the actual occupants at the base. After 1290, however, the ability to create new tenures was abolished.

Technically, therefore, land is owned by the Crown. Its subjects own not the physical soil but an *estate* in the land. An estate in this context has a special meaning. It is not the physical parkland and farms, but an abstract entity interposed between the *tenant* and the land. Tenant here means one who holds freehold land of a superior lord. Today all such tenants hold directly from the Crown. It is in this sense that joint owners of freehold property are called either joint tenants or tenants in common. The tenant is given specific rights and powers for a period of time. In the sixteenth century in *Walsingham's* case this principle was expressed as follows:

'The land is one thing and the estate is another thing, for an estate in the land is time in the land, or land for a time, and there are diversities of estate, which are no more than diversities of time.'

The estate indicates the period of time that the tenant enjoys in the land. He may have a *life interest* or a *fee simple*, that is an estate which lasts as long as he has heirs. Should a fee simple owner die without a will, or without relatives who are entitled to claim under the rules of intestacy, the property will revert to the Crown. This is the result of the Crown's existing property right as the superior lord, rather than its right to *bona vacantia*, the right it has to personal property which has no owner.

It is this splitting up of interests on a time basis which has enabled settlements to be made providing for land to be retained in the family for future generations. A settler might leave his property to his son for life, then to his son's wife and then to the children of the marriage with charges and annuities in favour of other members of the family. Taxation has gradually made these complex arrangements less attractive.

Despite the present fiscal disadvantages, the trust, which permits separation of the legal management of the land and the beneficial interests, is one of the great original contributions of English jurisprudence. European jurisdictions have been hampered by their more absolute approach to the physical ownership of immovable property. Without the abstract notion of the estate, fragmentation of interest amongst a number of people, each with a separate property right, is impossible.

Not only does a landowner have intangible rights over physical land, he may also have intangible rights over intangible land. For land means *corporeal* and *incorporeal hereditaments*. Incorporeal hereditaments include advowsons (the right of presentation to a living) titles of honour such as a peerage or lordship of the manor, easements (for example a right of way) and profits à prendre which include fishing and shooting rights.

It is therefore useful to see land ownership as a bundle of rights. What is in the bundle depends on what rights are vested in other people. The bank may have a secured loan (mortgage or charge), a neighbour may have a right to cross the field (easement) another may be able to restrict the building of a house on the land (restrictive covenant) or the land may be subject to a lease.

1.2 The legal estates

1.2.1 *The freehold*
The technical term is the *fee simple absolute in possession*. The word fee indicates that it is an estate of inheritance and simple that it can be inherited by the heirs generally and not a particular class of heir, although the owner of the fee simple can specify in his will who can inherit. Absolute distinguishes the fee from modified fees which today can only exist as equitable interests under a trust. Possession includes not only physical possession but also receipt of the rents and profits. Thus someone can have a fee simple absolute in possession even when the whole of the property is let to a tenant.

1.2.2 *The leasehold*
This is a term of years absolute, sometimes referred to as a *lease* and also a *tenancy*. It may be for a fixed term, for example ten years, or a periodic tenancy, such as a yearly, monthly or weekly tenancy. A lease is essentially

Introduction to Land Law

a contractual document. However, there has been much statutory intervention to protect residential, agricultural and other business tenants. The legislation which is most likely to be of concern to landowners and farmers is discussed in Chapters 2, 3, 4 and 5.

1.2.3 *The trust*

The freehold and leasehold estates are the only estates which can exist as *legal* estates. Any other estates can since 1925 only exist as an *equitable* interest under a *trust*. The legal estate will be vested in trustees who will hold the estate on trust for the beneficiaries. Since the Trusts of Land and Appointment of Trustees Act 1996 all new trusts involving land are trusts of land governed by that Act. Existing trusts for sale will become trusts of land. It will not be possible to create new strict settlements, except in very limited circumstances. Existing strict settlements will however continue. In a strict settlement, provided the life tenant is of full age and capacity, the legal estate will be vested in him but he will hold it as trustee according to the terms of the trust.

1.3 Legal interests

1.3.1 *The five legal interests*

Section 1 of the Law of Property Act 1925 provides that there are only five possible legal interests. These are

(1) Easements, rights and privileges
(2) Rent charges
(3) Mortgages
(4) Certain statutory charges
(5) Rights of re-entry

In order to be legal they must be granted formally for the equivalent of a freehold or leasehold interest.

The most important interests are easements and mortgages.

1.3.1.1 *Easements*

An *easement* is a private right enjoyed by successive owners over neighbouring land. There are four essential requirements

(1) There must be two plots of land. The burden must be attached to one plot of land (the servient tenement) for the benefit of another (the dominant tenement).
(2) The easement must improve the amenity of the dominant tenement.
(3) The dominant and servient tenement must be separately occupied. No one can have an easement over his own land.
(4) The right must be capable of being the subject matter of an easement.

An easement may be granted by Act of Parliament. For instance, electricity suppliers may have statutory rights to lay cables. Easements may also be expressly granted or reserved by deed. The extent of the right granted will depend on the terms of the deed.

Easements may be created by implication when an estate is divided. Rights which have been enjoyed over part of the estate may be automatically converted into full rights, under the rule in *Wheeldon* v. *Burrows* (1879) 12 Ch 31, or by virtue of section 62 of the Law of Property Act 1925, when the part is sold. In some exceptional cases there will be an implied reservation in favour of the vendor over his retained land. The court is less anxious to imply such rights because the vendor should have reserved them expressly and should not derogate from the grant he has made.

Easements may also arise where a right has been enjoyed without force, secrecy, or permission for at least 20 years. In *Hanning* v. *Top Deck Travel* (1993) 68 P&CR 14, it was held that no right could arise from long use where the use was contrary to statute.

Easements which are not granted for the equivalent of a legal estate, or are expressly granted but not by deed can exist as equitable interests. In order to bind a purchaser they will need to be protected by registration.

It is possible that an easement may be extinguished by agreement. Mere non-use will not cause an extinguishment. It is necessary to prove an intention to abandon the right.

The easements which are most usually encountered on farms are private rights of way and drainage rights.

1.3.1.2 Mortgages

A mortgage is a loan secured on property so that if the borrower defaults in his repayments the lender can sell the property and recover the money lent out of the proceeds of sale. The loan is normally repayable over a long term of years.

Mortgage deeds often provide that the property must not be let. By section 99 of the Law of Property Act 1925 the power of the borrower to grant a lease cannot be excluded in relation to any mortgage of agricultural land made after 1 March 1948 but before 1 September 1995. It is possible to preclude a borrower from granting a farm business tenancy under the Agricultural Tenancies Act 1995.

1.4 Equitable interests

1.4.1 Family equitable interests under a trust

Where there is joint ownership, successive interests or interests which will vest in possession in the future the legal estate will be held by trustees but the equitable interests will be in the beneficiaries. A life interest for example will be an equitable interest under a trust. Subject to the terms of

Introduction to Land Law

the trust deed, the trustees may sell the property. Provided the capital is paid to two trustees the purchaser will get a good title and will take free of the interests of the beneficiaries. The life tenant's interest is said to be *overreached*. He will be entitled to the income from the capital or any investment which replaces the land. On his death the capital will go to those who are absolutely entitled.

1.4.2 Commercial equitable interests

Easements, rent charges, mortgages and rights of re-entry can exist as equitable interests if they were not created in accordance with the formalities needed for creating a legal interest or if the grantor lacked either the intention or capacity to create a legal interest.

In addition there is a wide range of interests which can only be equitable. These include *estate contracts*, (that is contracts to convey or create a legal estate) *licences* and *restrictive covenants*.

Generally commercial equitable interests, unlike most legal interests, need protecting by registration if they are to bind a purchaser of the legal estate. This can be done as a land charge under the Land Charges Act 1972 where the title is still unregistered, or by notice or caution on the register under the Land Registration Acts. If they are not registered, a purchaser will take free of them, whether he knows of them or not.

There is a residuary class of equitable interests which can not be registered or overreached. These include licences by estoppel. Such an interest will bind a purchaser with notice.

1.4.3 Restrictive covenants

Restrictive covenants are among the more important equitable interests for a landowner. They can be imposed when part of a farm is sold and the sellor wants to control the use the purchaser makes of the land. A purchaser might, for example, covenant to use the land for agricultural purposes only.

At common law the burden of a covenant cannot pass with the land. Equity has intervened to allow negative covenants to pass with the land if certain conditions are fulfilled. The covenant must be negative in nature. It can be negative even if expressed positively. The test is 'does the covenant require the expenditure of money?' If it does, the covenant is positive. The person imposing the covenant must have other land for whose protection the covenant was taken. Such land must be capable of being benefited by the covenant. The covenant must have been intended to run with the land and not just bind the parties to the covenant. The covenant must have been protected by registration.

Provided these conditions are fulfilled the person who imposed the covenant can enforce it not only against the original covenantor but also his successors in title. The benefit of the covenant may be assigned to successors in title of the covenantee or may by the wording of the covenant be automatically annexed to the land.

It is possible to apply to the Lands Tribunal for the discharge of a restrictive covenant. The applicant must prove that the covenant is obsolete or obstructive to the use of the land or that the person entitled to the benefit of the covenant has agreed to, or will not be injured by, its discharge.

1.5 Registration of title

1.5.1 Compulsory registration

The purpose of registration of title is to replace the separate investigation of title, going back at least 15 years, on every purchase, by a title guaranteed by the state. The purchaser inspects the register to see whether the seller has power to sell and to discover the most important encumbrances affecting the property. The system was not meant to alter the substantive law but only the conveyancing machinery. However, it has become apparent that there are substantial differences depending on whether the title is registered or not.

Title to land must now be registered for the following transactions whether they are commercial transactions or gifts:

(1) A conveyance of the freehold
(2) A grant of a lease or underlease for more than 21 years
(3) An assignment of a lease or underlease having more than 21 years left to run
(4) A disposition by assent or vesting deed of a freehold or of a lease or underlease having more than 21 years left to run
(5) A grant of a legal mortgage protected by the deposit of title documents

1.5.2 Open register

Since 1990, when the Land Registration Act 1988 came into force, the land register has been open to public inspection. Although there was opposition by landowners to the register being made public there are many advantages. Amongst other things it is now possible to ascertain the ownership of unoccupied land and there is greater security of title when purchasing a leasehold interest.

1.5.3 Registered interests

The only estates which can be registered are legal estates. These are the fee simple absolute in possession and leases which at the time of first registration are granted for a term of more than 21 years or were assigned with more than 21 years to run.

A manor may be registered with its own title but the Crown cannot register land which it holds in its absolute ownership. This is because it is an estate in land which is registered and not the land itself. As explained at

Introduction to Land Law

section 1.1 above the Crown does not have an estate in land but owns the actual soil. The Land Registration Bill, currently before parliament, reverses the position. Lordships will no longer be registrable but Crown land will be.

1.5.4 Overriding interests

These rights do not appear on the register but bind a purchaser whether he knows of them or not. The most important of these interests are legal easements, rights being acquired by adverse possession under the Limitation Act 1980, chancel repair liability, leases not exceeding 21 years granted at a rent without a premium being taken, and in the case of possessory, qualified or good leasehold title all interests excepted from the effect of registration. The category of overriding interests which has caused the most difficulty is 'the rights of every person in actual occupation of the land or in receipt of the rent and profits thereof, save where enquiry is made of such person and the rights are not disclosed'.

This category includes tenants and those who go into occupation under an agreement for a lease. An option to purchase the reversion granted in a lease of registered land will bind a purchaser of the reversion as an overriding interest if the lessee is in occupation, even though the interest is not protected by a notice or caution on the register. Were the title unregistered such an interest would be void against the purchaser of the reversion if it was not protected by registration as a land charge.

If a beneficiary under a trust of land is in occupation, his interests will be protected. However the rights of a beneficiary under a settlement governed by the Settled Land Act 1925 are specifically excluded.

1.5.5 Minor interests

These rights will bind a purchaser only if they are protected by an entry on the register. There are two main categories of minor interest. The first category is equitable interests of beneficiaries under a trust of land or strict settlement. The second is equitable interests such as restrictive covenants which in unregistered land would need to be protected as a land charge under the Land Charges Act 1972 if they are to bind a purchaser.

Minor interests are protected by the entry on the register of a *notice*, *caution*, *inhibition* or *restriction*.

A notice is entered on the charges register, usually with the agreement of the registered proprietor. There are certain statutory provisions which provide for specified interests to be protected by notice, even where there is no agreement.

There are two kinds of caution. A caution against first registration may be lodged by any person with an interest in the land. The land registrar must then inform the cautioner of any application for the registration of the title. A caution against dealings protects minor interests where the registered proprietor is uncooperative and will not lodge his land certificate. The registrar must give notice to the cautioner before registering any dealing with the land. In both cases the cautioner has a fixed time after receiving notice, usually 14 days, to make his objections known. If he does not object the caution will be removed.

Cautions can protect such interests as options to purchase, equitable charges and easements. However, a caution gives only short-term protection because of the warning off procedure and the substantial discretionary powers given to the registrar, who is unwilling to allow cautions to be prolonged indefinitely. Clear titles are favoured at the expense of third party rights.

Inhibitions are orders of the court or the registrar which forbid dealing with the property either absolutely or until a certain time or event. They are only used where there is no other way of protecting a claim. An example is a bankruptcy inhibition preventing the registered proprietor disposing of his land where a bankruptcy order has been made.

Restrictions are entries made either by the registered proprietor or with his consent. They prevent any dealing with the land until there has been compliance with a specified condition. For example where there is a trust of land there is likely to be a restriction providing that there shall be no disposition unless capital money is paid to two trustees or the consent of specified persons is obtained. Restrictions are also used to ensure that on future dispositions the transferee enters into a direct covenant with the original vendor of the land.

1.5.6 Parts of the register

The register is divided into three parts. A copy of the entries is contained in the *land certificate* which is given to the registered proprietor (unless there is a mortgage when the borrower will be given a *charge certificate*).

The property register describes the registered land, refers to the general map or file plan and contains notes of interests held for the benefit of the land, such as easements and restrictive covenants.

The boundaries of the land are general boundaries only. There is a procedure for fixing boundaries but this is expensive and seldom used. Therefore in order to establish exact boundaries it may be necessary to study the pre-registration documents and to consider whether the boundaries have been altered over the years by adverse possession. If all else fails resort may be had to the boundary presumptions. These include the presumption that where properties are divided by a track, whether public or private, or a stream, the boundary is the centre line of the stream or track. A further presumption is that where there is a hedge and a ditch the boundary is the far side of the ditch from the hedge. This is based on the fanciful theory that a man digs a ditch on the boundary of his land, throws the earth over his shoulder onto his own land and on the mound so formed a hedge is planted. Of course this presumption will not apply where the land on both sides of the boundary was in the same ownership at the time the ditch was made. Where land is conveyed expressly by reference to an Ordnance Survey map the presumption is that the centre of a hedge will be the boundary.

The second part of the register is the *proprietorship register*. This gives the nature of the title (absolute, qualified, possessory or good leasehold). The classification depends on the approval of the registrar. The name, address and description of the proprietor is also given and any cautions, inhibitions or restrictions affecting his right to deal with the land are set out.

The third part, the *charges register*, contains entries of rights adverse to the land including mortgages, restrictive covenants and all notices protecting rights over the land.

1.5.7 Types of title

A freehold may be registered with absolute, possessory or qualified title.

The registration of a person with absolute title vests in him the legal estate subject to:

(1) incumbrances and other entries appearing on the register;
(2) overriding interests;
(3) where the proprietor is not holding for his own benefit minor interests of which he has notice.

So a trustee would be subject to the equitable interests of the beneficiaries.

Although absolute title is the best known to English law it is still only relative. Overriding interests are a major blot on the title. Moreover in certain circumstances the register may be rectified.

Possessory title has the same effect as absolute title except that registration does not prejudice the enforcement of any adverse interest subsisting at the date of first registration. There is no guarantee in respect of the prior title before registration.

Qualified title also has the same effect as absolute title except that the property is held subject to some defect or right specified on the register. It is very rare in practice to find a qualified title.

Leaseholds can be registered with absolute, possessory or qualified title in the same way as freeholds. Absolute title will only be registered where the freehold and any intermediate titles have been registered. Absolute leasehold title is a guarantee that the proprietor is the owner of the lease and that the lease has been properly granted.

As the aim of registration of title is to facilitate the transfer of good title, power is given to the registrar to upgrade the category of title. The registrar may convert

- a good leasehold title to absolute title if he is satisfied as to the title to the freehold and to any intermediate title;
- a freehold possessory title to an absolute title and a possessory leasehold to good leasehold, if he is satisfied as to the title, or if the land has

been registered with possessory title for at least 12 years and he is satisfied that the proprietor is in possession;
- a qualified freehold title to absolute title and a qualified leasehold title to good leasehold if he is satisfied as to title.

1.5.8 Rectification

Even where the proprietor has been registered with absolute title the register may be rectified against him. Rectification may take place to give effect to a court order. The Land Registration Act 1925 and the Land Registration Rules 1925 allow rectification on a number of different grounds. These include entries or omissions made by fraud, mistake or other errors. However, the register will not be rectified against a registered proprietor in possession unless

- it is to give effect to an overriding interest;
- the proprietor has caused or substantially contributed to the error by fraud or lack of proper care;
- it is to give effect to a court order; or
- for any reason it would be unjust not to rectify against him.

Even though an applicant is able to bring himself within one of the cases where rectification may be claimed against a registered proprietor in possession, the jurisdiction is still discretionary. In assessing justice the court can take into account the fact that rectification would entitle the losing party to indemnity while non-rectification would not. Rectification may also be refused if the indemnity will not be adequate compensation for the loss of the land.

1.5.9 Indemnity

The Land Registration Act 1925 gives a right of indemnity to persons who suffer loss where

(1) The register is rectified
(2) There is an error or omission but the register is not rectified
(3) Documents lodged at the registry are lost or destroyed or there is an error in an official certificate of search
(4) Rectification affects a proprietor claiming in good faith under a forged disposition

Indemnity for non-rectification is limited to the value of the lost interest at the time when the mistake was made. Where the register is rectified the indemnity will be the value of the lost interest immediately before the time of rectification. A claim to indemnity is barred after six years from the time when the claimant knew, or should have known of the existence of his claim.

No indemnity will be paid where the applicant has caused or substantially contributed to the loss by fraud or lack of proper care. Further, there must be a loss so no indemnity will be given where rectification is to give effect to an overriding interest. A purchaser takes subject to overriding interests so rectification is only recognising the existing position.

It is often said that under the system of registration the state guarantees the title. This is an exaggerated claim. A transferee takes subject to overriding interests which do not appear on the register. The court has a wide discretion to rectify the register as may seem just, and there are a number of situations where the true owner or innocent purchaser may be left without property or compensation.

Chapter 2
Farm Business Tenancies

2.1 Introduction

The Agricultural Tenancies Act 1995 came into force on 1 September 1995. It applies to both oral and written farm business tenancies beginning after that date. The Agricultural Holdings Act 1986 will continue to apply to existing tenancies and succession tenancies.

The underlying principle of the Act is freedom of contract with maximum flexibility to enable the parties to enter into agreements which suit their own individual needs. No security of tenure is given to the tenant other than that provided by the contract itself. There are very few mandatory provisions. These relate to

(1) Definition of a farm business tenancy
(2) Termination of a farm business tenancy
(3) Rent
(4) Fixtures and compensation
(5) Resolution of disputes

2.2 Meaning of farm business tenancy

The Act defines the term 'farm business tenancy'. In order to be a farm business tenancy, the tenancy must comply with the business conditions and either the agriculture or notice conditions (section 1 Agricultural Tenancies Act 1995).

2.2.1 Business conditions
There are two business conditions. First, since the beginning of the tenancy, all or part of the land must have been farmed for the purpose of a trade or business. 'Farmed' is defined in section 38(2) of the Act to include references to the carrying on in relation to land of any agricultural activity. The definition of 'agricultural' is similar to that in the Agricultural Holdings Act 1986. Both definitions use the word 'include' and are not therefore exhaustive. It is considered that the expression 'farming of land', because it includes agricultural activity, may be wider than 'agriculture'. The other condition is that, at the time when the status of the tenancy is challenged, all or part of the land comprised in the tenancy must be so farmed. Provided this second condition is fulfilled at the time of challenge then it is presumed that the first condition has been fulfilled unless the contrary is proved. This presumption may be of assistance where proceedings have arisen in respect of another issue but while that issue is

being determined one of the parties questions whether the business condition has been fulfilled at all times since the tenancy was granted. This could be difficult to prove and therefore without the presumption would impose an unreasonable burden on the party challenged.

It is not necessary for the same part of a holding always to have been farmed for the purpose of a trade or business. The business conditions will be satisfied if farming for a trade or business has been carried out on different areas of the holding during the tenancy, provided that at all times there has been some part of the holding in use for commercial farming. The business condition does not require the farming activity to be predominant. Any farming activity will suffice provided it is not *de minimis*.

If all agricultural activity ceased then the tenancy would no longer be a farm business tenancy under the Agricultural Tenancies Act 1995. It would become a business tenancy under the Landlord and Tenant Act 1954, Part II. There would be no opportunity for the parties to make a joint application to the county court to contract out of the security provisions of the 1954 Act. This is because such an application has to be made before the tenancy is granted. A tenancy which ceased to be a farm business tenancy automatically would become a business tenancy without a further grant.

The Landlord and Tenant Act 1954, Part II applies

> 'to any tenancy where the property comprised in the tenancy is or includes premises which are occupied by the tenant and are so occupied for the purposes of a business carried on by him or for those and other purposes.' (section 23(1))

Business is widely defined. It includes

> 'a trade, professional or employment and includes any activity carried on by a body of persons, whether corporate or incorporate.' (section 23(2))

Therefore, although land let for non-commercial farming will not be a farm business tenancy, it could be a business tenancy. Premises for the purposes of the 1954 Act include bare land. Agricultural tenancies and farm business tenancies are expressly excluded from the Landlord and Tenant Act 1954.

Where the tenancy becomes a business tenancy under the 1954 Act either because it fails to comply with the business conditions or the agriculture condition (discussed in section 2.2.2 below) then the tenant has a right to apply to the court for a new tenancy at the end of his lease. However, a landlord can oppose the grant of a new tenancy on the ground that he intends to occupy the premises for the purpose of a business carried on by himself (or by a company which he controls) or as his residence.

If the farming activity does not amount to a business then the tenancy will simply be a common law tenancy without any statutory protection. It will not be governed by the Agricultural Tenancies Act 1995 nor by the Landlord and Tenant Act 1954.

2.2.2 *The agriculture condition*

The agriculture condition, unlike the business condition, does not need to have been satisfied at all times since the tenancy was granted. It must be satisfied only at the date of challenge. The agriculture condition is that, having regard to

- the terms of the tenancy;
- the use of the land comprised in the tenancy;
- the nature of any commercial activities carried on on that land; and
- any other relevant circumstances,

the character of the tenancy is wholly or primarily agricultural.

There is no single test or any weighting given to the elements which make up the condition. One criterion, taken alone, could give a misleading answer as to the overall character of the holding.

The drafting of the condition enables the court to use reasonable discretion in judging the character of the tenancy, by having regard to the range of factors specified. The terms of the tenancy will not necessarily be in writing and there may be a dispute over what oral terms and conditions were negotiated by the parties. The terms of the tenancy would then be a matter for the court to determine on the evidence. Where there is a written agreement, the terms of the agreement will be relevant in deciding the character of the tenancy.

If the status of a tenancy is challenged early in the tenancy, discretion may need to be exercised to determine what use is being made of the holding. For example, non-agricultural crops may have been sown on certain areas within a holding or non-agricultural use may take place for part of a year. It may be that at the time of the challenge there is no financial evidence available (or the evidence may not cover a reasonably representative period of time) to enable the respective contributions to the business of the agricultural and commercial activities to be assessed. For instance, in a given year, income from agriculture might be virtually non-existent and thus outweighed by only modest non-agricultural revenue.

The judge or arbitrator will have to consider all these matters and any additional relevant circumstances, including allowance for exceptional circumstances, in determining whether or not the agricultural condition is fulfilled at the date of the challenge.

2.2.3 *Notice conditions*

The purpose of these conditions is to ensure that the parties, especially the tenant, are aware of the nature of the tenancy before they become contractually bound. However, failure to give notice will not prevent the tenancy being a farm business tenancy provided the tenancy complies with the business and agriculture conditions. The notice is not therefore so crucial as the notice which has to be given before the grant of an assured shorthold tenancy of a dwelling.

Nevertheless, it will be more satisfactory if the parties do in fact comply with the notice conditions. This will ensure that the tenancy will remain a farm business tenancy and therefore within the legislation as intended and contemplated by the parties. If the notice conditions are not fulfilled then there is a danger that on the development of a non-agricultural business by the tenant, the tenancy will cease to be a farm business tenancy and become a business tenancy governed by the Landlord and Tenant Act 1954. Well advised landlords and tenants will ensure therefore that there is compliance with the notice conditions, to avoid uncertainty. In practice, it is likely that where there are written tenancies the notice conditions will be fulfilled but where there are more informal arrangements, resulting in oral tenancies they will not. Also, where there are fixed term tenancies of two years or less with a covenant restricting the use to agriculture, the parties may decide to dispense with notices.

There are two conditions which must be fulfilled. On or before the day upon which the parties enter into any instrument creating the tenancy, other than an agreement to enter into a tenancy on a future date, or before the beginning of the tenancy, if this is earlier, the parties must give each other a written notice which

- identifies the land to be comprised in the tenancy whether by name or in some other way (such as Ordnance Survey parcel reference numbers, or by means of an attached plan); and
- contains a statement to the effect that the person giving the notice intends that the tenancy is to be and remain a farm business tenancy throughout the term.

The notice does not have to be in a prescribed form and there is no need for it to be signed. The intention of the legislation is that the notice procedure should be kept as simple as possible. However, the notice cannot be included in the tenancy agreement itself.

The second condition is that at the beginning of the tenancy the character of the tenancy must be primarily or wholly agricultural. In deciding this question, regard has to be paid to the terms of the tenancy and any relevant circumstances. There is no obligation to have regard to the use of the land comprised in the tenancy or the nature of any commercial activities on the holding, these being the other two factors to be taken into account when assessing whether the agriculture condition has been fulfilled. This is because the notice condition applies at the beginning of the farm business tenancy and there may be no valid evidence at that time regarding those matters which could be applied to an incoming tenant. However, where the tenancy is a new tenancy, but granted to the same tenant on the same terms and conditions as the tenancy which immediately preceded it, there may be evidence about the use of the relevant circumstances which have to be taken into account in deciding whether the use was primarily or wholly agricultural at the start of the tenancy.

Provided that the notice condition has been fulfilled then it does not matter that there is a subsequent shift in emphasis of activity away from agriculture.

It should be stressed, though, that the use must be primarily or wholly agricultural at the beginning. So a tenancy granted for a riding school, even if it included grazing land would not qualify as a farm business tenancy but would be a business tenancy governed by the Landlord and Tenant Act 1954.

The 1995 Act sets out circumstances in which the notice conditions will be deemed to have been complied with, where a new tenancy follows immediately upon a previous tenancy which met the notice condition. The effect is that where there is a surrender and regrant of a farm business tenancy in respect of which notices were exchanged before that tenancy was granted, the parties do not have to exchange further notices in respect of the new tenancy.

The deeming provision applies to both express and implied surrenders and regrants. An implied surrender and regrant is most likely to occur where there is a minor change in area of a holding which is otherwise occupied under identical terms to those contained in the original tenancy. The parties do not always appreciate that in such circumstances a new tenancy has been created and might not think of serving new notices.

The rule that no new notice need be served applies only where the following conditions are met.

(1) Notices must have been exchanged by the landlord and tenant in respect of the original tenancy.
(2) The new tenancy must be between the same landlord and the same tenant.
(3) The terms of the new tenancy must be substantially the same as the old except
    (a) for changes in area which are small in relation to the size of the holding and do not affect the character of the holding; or
    (b) the only difference is that the new tenancy is for a fixed term which expires earlier than the fixed term under the old tenancy.

Consequential changes or change in area, or reduction in term, agreed by the parties are permitted. An example might be a small increase in rent to cover a small increase in the area of the holding.

If, on the other hand, there is a substantial alteration in the terms, or the new tenancy is for a longer fixed term than the original tenancy, new notices must be exchanged. New notices will also be necessary where there is a series of short-term tenancies which expire by effluxion of time.

2.2.4 Effect of breach of covenant

A tenant who is in breach of any of the terms of his tenancy as to the use of the land, the commercial activities or cessation of commercial activities on the holding, cannot challenge either the business conditions or the agriculture conditions by relying on his own breach. For example, a tenancy agreement might prohibit the tenant from carrying on any commercial activities other than agriculture on the holding. If the tenant began such activities and sought to challenge the agriculture condition on the basis that the holding was no longer primarily or wholly agricultural in character, the court would disregard the effect of the breach when applying the agriculture condition. There might also be a requirement in a tenancy agreement that part of the land would at all times be farmed for the purposes of a trade or business. A breach of this requirement would be disregarded when applying the business conditions. However, the position would be different where the landlord, or a previous landlord, had consented to or acquiesced in the breach. In those circumstances, the breach would not be disregarded in applying the business conditions or the agriculture condition.

2.3 Termination of a farm business tenancy

2.3.1 Fixed term tenancies

Fixed term tenancies of two years or under expire automatically on the term date. For a tenancy over two years, a written notice of at least 12 but less than 24 months must be served expiring on the contractual termination date (section 5 Agricultural Tenancies Act 1995). If no notice is served, then the tenancy will continue as a yearly tenancy on the terms of the original tenancy so far as these are applicable to a tenancy from year to year. The yearly tenancy can be terminated by a written notice of at least 12 but less than 24 months expiring at the end of the completed year. The notice can be served during the last year of the fixed term so that the tenancy is terminated at the end of the following year. It is not possible to contract out of these provisions (section 6 Agricultural Tenancies Act 1995).

2.3.2 Periodic tenancies

In order to terminate a yearly tenancy a written notice of at least 12 but less than 24 months taking effect at the end of a year of the tenancy must be served. It is not possible to contract out of this provision (section 6 Agricultural Tenancies Act 1995). A yearly tenancy will continue after the death of either party until notice of between one to two years is served.

Other periodic tenancies are subject to the normal common law rules. This means that a tenancy from week to week, month to month or quarter to quarter must be terminated by notice of the full period expiring at the end of the completed period unless the parties agree otherwise. There is no provision in the Agricultural Tenancies Act 1995 dealing with periodic tenancies other than those from year to year.

Where there is a dwellinghouse comprised in the letting, then at least four weeks notice must be given, even though the tenancy may be a weekly tenancy (section 5 Protection from Eviction Act 1977 as amended by sections 25–32 Housing Act 1988).

2.3.3 Break clauses
A farm business tenancy may contain a break clause giving either the landlord or the tenant the option of terminating the tenancy or the holding or part of the holding (section 7 Agricultural Tenancies Act 1995). Where the tenancy is for more than two years, this option can be exercised only where a notice has been served of at least 12 but less than 24 months before the date when it is to take effect. It is not possible to contract out of this statutory requirement.

2.3.4 Severed reversion
Where there is a right to serve a notice to quit and the landlord has assigned part of the reversion, any landlord of the severed part can serve a notice to quit relating to that part (section 140 Law of Property Act 1925). However, the tenant is given the option of quitting the entire holding by serving a counter notice within one month upon the reversioner(s) of the rest of the land. The notice given by the landlord of the severed reversion will have to comply with the notice rules relating to a fixed term tenancy or a tenancy from year to year, or to the exercise of an option to terminate the tenancy or resume possession of part. In all three situations the notice must be given between 12 and 24 months before the date on which it is to take effect. On the other hand, the counter notices served by a tenant in response do not have to comply with the statutory rules in the 1995 Act for notices terminating yearly tenancies and exercising break clauses (sections 6 and 7 Agricultural Tenancies Act 1925).

2.3.5 Lease for lives
A lease at a rent granted for life or lives or for a term of years determinable with a life or lives or on the marriage of the lessee is converted into a term of 90 years (section 149(6) Law of Property Act 1925). The lease continues after the death or marriage but can then be determined by either party serving on the other one month's written notice expiring on one of the quarter days applicable to the tenancy, or, if there are no special quarter days on one of the usual quarter days. Such leases are extremely rare today. But the 1995 Act provides that the one to two year notice provision will not apply (section 7(3) Agricultural Tenancies Act 1995). There was a similar exception under the Agricultural Holdings Act 1986.

2.3.6 Other methods of determination
Besides the special provisions in the Agricultural Tenancies Act 1995 for

terminating a tenancy by notice, tenancies may be brought to an end by an agreement to surrender or by merger where the tenant acquires the interests of the landlord or where a third party acquires both the lease and the reversion.

If there is a forfeiture clause in the lease, then the landlord may forfeit on breach of covenants provided he complies with the normal rules for forfeiture.

It should be noted that there are no provisions enabling the landlord to serve an incontestable notice to quit as there is under the Agricultural Holdings Act 1986 (see Chapter 3, section 3.6.4).

2.4 Rent

2.4.1 *Principles*
The underlying principle of the legislation, namely freedom of contract, is reflected in the provisions on rent. The parties are free to agree the initial rent, whether or not this will be reviewable and, if so, the dates and frequencies of any such reviews. However, if the contract makes provision for an upward only rent review, then the tenant can demand that the matter be referred to arbitration. The arbitrator can determine the rent only on an open market basis.

2.4.2 *Fixed rent*
The parties may agree that the rent will remain the same throughout the term (section 9(a) Agricultural Tenancies Act 1995). It is essential that the contract expressly states that the rent is not to be reviewed during the tenancy. Failure to do this will mean that the statutory rent review provisions will come into effect. Any agreement which is not set out in a written tenancy agreement will not be effective.

2.4.3 *Phased rent*
Parties may also agree that the rent will be increased at a specified time to or by a specified amount (section 9(b)(1) Agricultural Tenancies Act 1995). The tenancy agreement must indicate that apart from such specified changes, the rent is to remain fixed.

2.4.4 *Objective criteria*
It is also possible for the parties to stipulate in a tenancy agreement that the rent shall be adjusted by reference to a formula which does not require or permit the exercise by any person of any judgement or discretion in relation to the determination of the rent of the holding (section 9(b)(ii) Agricultural Tenancies Act 1995). The formula must not be so drafted as to preclude a decrease in rent. It is not possible to have what is, in effect, an upwards only rent review. Again it is necessary for the tenancy agreement to state clearly that apart from the specific terms of variation according to

the agreed criteria, the rent is to remain fixed. This is to ensure that the statutory rent review provisions do not come into play.

2.4.5 Reviewable rents

The parties can agree that the rent will be reviewed. They can stipulate their own formula for determining the rent at a review date. If they do so, the agreement will be binding unless either the landlord or the tenant serves on the other a statutory review notice of at least 12 but not more than 24 months before the contractual review date (section 10 Agricultural Tenancies Act 1995).

2.4.6 Dates for rent reviews

Whether the rent is reviewed by reference to objective criteria or under contractual or statutory rent review provisions the parties can agree the dates for the rent reviews. The parties could specify the dates and the years for such reviews, or that the rent should be reviewed at specified intervals on a given date.

If no dates are agreed for a rent review and a notice requiring a statutory rent review has been served, then the review date must be an anniversary of the beginning of the tenancy.

2.4.7 Frequency of review

The parties can stipulate the frequency of rent reviews which may be more or less than the three years statutory fallback provision. Where there is no contractual provision, the statute gives the right for either party to demand a rent review every three years but not at shorter intervals.

The rent review period runs from either

'(i)   the beginning of the tenancy; or
(ii) the date from which there took effect a previous direction of an arbitrator as to the amount of rent; or
(iii) a date from which there took effect a previous determination as to the amount of rent made, other than as an arbitrator, by a person appointed under an agreement between the landlord and tenant; or
(iv) a date from which there took effect a previous written agreement between the parties as to the amount of the rent, provided that such an agreement was entered into after the grant of the tenancy.'

section 10 Agricultural Tenancies Act 1995

2.4.8 Severed reversion

Where the reversion is severed, this has no effect on the tenancy which remains a single tenancy (*Jelley* v. *Backman* [1974] QB 488). However, it is possible for the landlord to agree with the tenant that there should be separate tenancies of the land comprised in the severed parts of the reversion.

Provided that

(1) the new landlord was immediately before the grant of the new tenancy entitled to a severed part of the reversionary interest in the original holding;
(2) the tenant was the tenant of that original holding; and
(3) the rent payable under the new tenancy in respect of the severed part is merely an appropriate proportion of the rent payable for the entire holding immediately before the new tenancy began,

then the minimum period of three years between statutory rent reviews is not affected by the creation of a new tenancy at a proportionate rent following severance of the reversion (section 11 Agricultural Tenancies Act 1995). The parties do not therefore have to wait a further three years before a rent review. Once a rent review has taken place under the new tenancy, it will have its own three yearly rent review cycle.

These provisions will not apply where the parties have agreed in writing on the frequency of rent reviews.

2.4.9 Amount of rent

The parties can contract that the rent

(1) shall remain the same throughout the tenancy;
(2) shall be increased to specified amounts at specified intervals; or
(3) shall be increased or decreased by reference to objective criteria.

If the rent is reviewed then the rent is determined at the review date (section 10 Agricultural Tenancies Act 1995).

The arbitrator must determine the open market rent at which the holding might reasonably be expected to be let by a willing landlord to a willing tenant and should take into account all relevant factors including, in particular, the terms of the tenancy (section 13 Agricultural Tenancies Act 1995).

The arbitrator is directed to disregard the fact that the tenant is in occupation. He must also disregard any dilapidation, deterioration or damage to the buildings or land caused or permitted by the tenant.

In general, rent will not be charged on any increase in the rental value of the holding that is attributable to the improvements made by the tenant. There are three exceptions when the increase value will not be so disregarded.

(1) Where there is an obligation on the tenant imposed by the tenancy agreement, or any previous tenancy agreement, to provide the improvement.
(2) Where the landlord has made the tenant any allowance or given any benefit in respect of the improvement, the arbitrator must take into

account the value of that allowance or benefit in determining the increase in rental value to the holding attributable to the improvement.

(3) Where the tenant has received compensation from the landlord for an improvement, the arbitrator must take into account that payment in determining the value of the holding attributable to the improvement.

2.4.10 Statutory review notice

Either the landlord or the tenant may serve a notice in writing on the other party requiring the rent for the holding to be referred to arbitration under the provisions of Part II of the Agricultural Tenancies Act 1995 (section 10 Agricultural Tenancies Act 1995).

The notice must specify the date from which the rent review is to take effect and that date must be at least 12 but less than 24 months after the day on which the notice was given.

The review date must be in accordance with any written agreement between the parties. The parties may have agreed in writing that the rent can be varied as from a specified date or dates at specified intervals during the tenancy. They may have agreed that the rent review date is a specified date or dates.

Where there are no contractual agreements, then the review date will be the anniversary of the tenancy. If the parties have not agreed on the frequency of reviews, then reviews cannot take place more frequently than every three years.

2.4.11 Appointment of an arbitrator

Once the statutory review notice has been served, the landlord and tenant may agree upon the rent, or on an arbitrator or some other person to determine the rent. The arbitrator can determine the rent only on an open market basis (section 13 Agricultural Tenancies Act 1995). On the other hand, a person appointed by agreement after the service of the statutory review notice, who is not acting as an arbitrator, can determine the rent on a basis agreed by the parties. If neither an arbitrator nor any other person is agreed upon, then either party can apply to the President of the Royal Institution of Chartered Surveyors for the appointment by him of an arbitrator (section 12 Agricultural Tenancies Act 1995).

Application to the President of the RICS may be made at any time during the six months before the review date, including the review date itself. If the rent has not been agreed and the parties have not appointed an arbitrator before the review date, a unilateral application can be made to the President. The application must be in writing and must be accompanied by such reasonable fee as the President may charge for the service.

If after a statutory review notice has been given, the parties agree on an arbitrator or other person to determine the rent, then an application cannot be made to the President for an appointment of an arbitrator by him.

2.5 Fixtures and compensation

2.5.1 Mandatory provisions
One of the purposes of the Agricultural Tenancies Act 1995 is to ensure that a tenant receives proper compensation for his investment in the business. The compensation provisions are mandatory and generally speaking the tenant will be able to opt to remove an improvement as a fixture rather than claiming compensation.

2.5.2 Fixtures
A farm business tenant has the right to remove any fixture, not just agricultural fixtures, which the tenant has affixed to the holding under the current tenancy or previous tenancies (section 8 Agricultural Tenancies Act 1995). The right contained in the 1995 Act is exhaustive. There is no additional common law right to remove trade or ornamental fixtures. He is also entitled to remove buildings and any part of a building.

The right of removal must be exercised at any time during the course of the tenancy. It may also be exercised after the tenancy has ended, provided that the tenant has not actually quit the holding but remains in possession as tenant.

The tenant must not cause any avoidable damage to the holding during the removal of the building or fixture. Any damage done must be made good immediately.

The tenant has the right to remove fixtures or buildings acquired by him, as well as those fixed or erected by the tenant himself.

The tenant cannot remove buildings and fixtures where

(1) the tenant was required, either by the tenancy agreement or otherwise, to affix a fixture to the holding or to erect a building;
(2) a tenant substitutes a fixtures or building for one belonging to the landlord;
(3) compensation has already been paid to the tenant for an improvement made to the holding;
(4) the landlord has given consent under section 17 of the 1995 Act to the tenant making the improvement on the condition that the improvement is not removed and the tenant has agreed to that condition.

2.5.3 Compensation for improvements
A tenant is entitled to compensation for improvements which he has made with the landlord's written consent (section 16(1) Agricultural Tenancies Act 1995). The improvement must have been made at the expense, either

wholly or partially, of the tenant and must not be removed from the holding at the end of the tenancy. The improvement must add to the letting value of the holding.

2.5.4 Definition of an improvement
The tenant's improvement is either

'(a) a physical improvement made on the holding by the tenant by his own effort or wholly or partly at his own expense; or
(b) an intangible advantage which –
 (i) is obtained for the holding by the tenant by his own effort or wholly or partly at his own expense; and
 (ii) becomes attached to the building.'
 (section 15 Agricultural Tenancies Act 1995)

There is a sub-definition of routine improvements. This is an improvement which (1) is a physical improvement made in the normal course of farming the holding or any part of the holding and (2) does not consist of fixed equipment or an improvement to fixed equipment but does not include any improvement whose provision is prohibited by the terms of the tenancy (section 19(10) Agricultural Tenancies Act 1995).

2.5.5 Right to compensation
In order to be eligible for compensation the tenant must have obtained the written consent of the landlord or that of the arbitrator to make the improvement (section 17 Agricultural Tenancies Act 1995). The landlord can give the consent before or after the improvement is made. However, for improvements, other than routine improvements, the tenant can only apply to the arbitrator, where the landlord has refused consent, if he has not begun to provide the improvement.

The consent may be given conditionally or unconditionally. Where the consent is given conditionally, the conditions must relate to the improvement and not to the amount of compensation. It is not possible to contract out of the compensation provisions.

The method of assessment of compensation is mandatory and cannot be altered by agreement. The value of the improvement cannot be written down.

2.5.6 Landlord's consent for planning permission
In order for the tenant to obtain compensation for any planning permission which he has secured, the following conditions must be satisfied:

'(a) the landlord has given consent in writing to the making of an application for planning permission,
(b) that consent is expressed to be given for the purpose
 (i) enabling a specified physical improvement lawfully to be made on the holding by the tenant, or

(ii) enabling the tenant lawfully to undertake a specified change of use, and
(c) on the termination of the tenancy the physical improvement has not been completed or the specified change of use has not been effected.'

(section 18(1) Agricultural Tenancies Act 1995)

The landlord has an absolute veto. Should he refuse to give his consent the tenant has no appeal to an arbitrator.

2.5.7 Application to arbitrator

For improvements, other than planning permission, a tenant may give to his landlord a notice in writing demanding arbitration (section 19(1) Agricultural Tenancies Act 1995) where he is aggrieved because his landlord

- has refused to give consent to an improvement; or
- has failed to give consent to the improvement within two months of the tenant's written request for such consent; or
- has required the tenant to agree a variation in the terms of the tenancy as a condition of giving such consent.

Where a landlord has refused consent, or imposed conditions on his consent, notice requiring arbitration must be given within two months beginning with the day on which notice from the landlord, refusing consent or requiring the variation of the tenancy as a condition of consent, was given to the tenant.

Where a landlord has failed to give a requested consent within two months of the tenant's written request to the landlord for such consent, notice requiring arbitration must be given within a further two months.

Once a tenant has given notice, the parties may agree who shall be appointed an arbitrator. Where no agreement is reached, either party may apply to the President of the Royal Institution of Chartered Surveyors for the appointment of an arbitrator by him (section 19 Agricultural Tenancies Act 1995).

2.5.8 Powers and duties of an arbitrator

The arbitrator must consider whether it is reasonable for the tenant to provide the proposed improvement, having regard to the terms of the tenancy. The arbitrator is also directed to take into account all other relevant circumstances including the circumstances, of the landlord and tenant.

The arbitrator may give or withhold his approval but he cannot give approval subject to conditions, whether his own or those of the landlord. Nor can he vary any conditions required by the landlord in respect of the landlord's consent.

2.5.9 Amount of compensation

The amount of compensation payable to a tenant for a tenant's improvement (other than for any planning permission obtained by the tenant), is an amount equal to the increase attributable to the improvement in the value of the holding at the termination of the tenancy as land comprised in a tenancy (section 20 Agricultural Tenancies Act 1995).

Deductions from the amount of compensation payable will be made where the tenant has been given an allowance or benefit for the improvement or a grant has been made from public money.

Where a tenant obtains planning permission but by the end of the tenancy the tenant has not made the physical improvement or effected the change of use specified by the landlord when giving his consent to the tenant applying for planning permission, compensation will be payable to the tenant. Compensation is based on the increase in value at the end of the tenancy of land comprised in the tenancy which is attributable to the fact that the specified physical improvement or change of use is authorised by planning permission (section 21 Agricultural Tenancies Act 1995).

If the parties have agreed in writing that the landlord will make a financial or material contribution towards the obtaining of planning permission, any compensation otherwise payable at the end of the tenancy in respect of the permission will be reduced proportionately.

2.5.10 Settlement of claims

If the parties are unable to agree on the amount of compensation, the tenant may within two months from the end of his tenancy, give notice in writing to his landlord of his intention to make a claim. The notice must comply strictly with the statutory time limit and must specify the claim (section 22(1) and (2) Agricultural Tenancies Act 1995).

After a notice has been given, the parties may still settle the claim by agreement in writing or they may appoint an arbitrator. Upon failure to do either of these things, either party may apply to the President of the Royal Institution of Chartered Surveyors for the appointment of an arbitrator. The application must be in writing and must be accompanied by such reasonable fee as the President may prescribe. An application cannot be made earlier than four months from the date on which the tenancy ended.

2.5.11 Successive tenancies

Where the tenant remains in possession under a new tenancy, he can roll over his right to compensation for improvements into the new tenancy (section 23 Agricultural Tenancies Act 1995).

2.5.12 Resumption of part of the holding

Where the landlord resumes possession of part of a holding, either because there is a contractual provision allowing him to do so, or there has been a severance of the reversion, that part of the holding is deemed to be a separate holding. Compensation is calculated as at the date on which possession is taken. The basis on which the compensation is assessed is the increase in value of the entire original holding attributable to improvements made on the part of the holding which is taken from the tenant. The 'original holding' means the land comprised in the farm business tenancy on the date that the landlord gave his consent to the improvement, or the date when an arbitrator gave his approval to the improvement (section 24 Agricultural Tenancies Act 1995).

Where the improvements are not on the land taken, but on the land retained, compensation is again calculated on the value which the improvement has on the entire original holding. The date for ascertaining the value, though, is different. It is the date of the termination of the tenancy.

2.5.13 *Compensation where reversion severed*

Even though the reversion may be severed, the tenant, on quitting the entire holding, can require that he be paid compensation as if there were a single landlord. This applies only to the statutory compensation to which the tenant is entitled (section 25 Agricultural Tenancies Act 1995). It will not apply to any additional contractual compensation which the parties may have agreed.

Where there has been a severance and the compensation is determined at arbitration, rather than by agreement, the arbitrator must apportion the liability for compensation between the persons in whom the reversionary interest is vested. Any additional cost to which such an apportionment gives rise are to be paid by such persons in such proportions as the arbitrator directs.

2.6 Dispute resolution

2.6.1 *Purpose*

The purpose of the dispute resolution provisions is to give the landlord and tenant the option of specifying their own dispute resolution procedures except for the determination of rent in pursuance of a statutory review notice, consent to improvements and compensation, when arbitration is compulsory. In addition to the statutory measures for settling disputes the parties can agree to some form of alternative dispute resolution procedure. The essence of such procedures is that it is open to the parties during the negotiations to decide to go to arbitration, or where arbitration is not compulsory, the courts.

2.6.2 *Determination of rent*

The landlord or tenant under a farm business tenancy may give the other a statutory review notice requiring that the determination of the rent from the review date shall be referred to arbitration (section 10 Agricultural

Tenancies Act 1995). The parties can then appoint an arbitrator by agreement or a third party, who is not acting as an arbitrator, to determine the rent.

2.6.3 *Consent to improvements*
In order for a tenant to be entitled to compensation for improvements at the end of his lease, he must have obtained the prior consent of his landlord (section 17 Agricultural Tenancies Act 1995). Where the landlord refuses, or fails to give consent, or imposes variations in the terms of the tenancy which are unacceptable to the tenant as a condition of giving his consent, the tenant may, by notice in writing given to the landlord, demand that the question should be referred to arbitration (section 19 Agricultural Tenancies Act 1995).

If the landlord refuses consent, or gives it subject to unsatisfactory conditions, the tenant must serve his notice within two months of receiving the landlord's refusal or conditions to consent. Where the landlord fails to give consent within two months of the tenant's request, the tenant must serve his notice within four months from his original request.

2.6.4 *Compensation*
Part III of the Agricultural Tenancies Act 1995 sets out the conditions which must be fulfilled before the tenant is entitled to compensation. Claims for compensation must be made in writing by the tenant within two months of the determination of the tenancy (section 22(2) Agricultural Tenancies Act 1995). Disputes about compensation are determined by arbitration.

2.6.5 *Failure to agree*
If the parties fail to agree on the rent, consent to improvements or compensation and do not appoint an arbitrator by agreement then either the landlord or tenant can apply to the President of the Royal Institution of Chartered Surveyors for the appointment of an arbitrator.

Time limits are as follows:

(1) For determination of rent – a period of six months ending with the review date (section 12 Agricultural Tenancies Act 1995)
(2) For consent to improvements – no time limit (section 19 Agricultural Tenancies Act 1995)
(3) Compensation – four months after the termination of the tenancy (section 22(3) Agricultural Tenancies Act 1995)

2.6.6 *Resolution by third parties*
The parties can also provide in their tenancy agreement that an independent third party can settle disputes rather than an arbitrator. The third party, whether or not acting as arbitrator, must be empowered by the

terms of the tenancy to give a decision which is binding in law on the landlord and tenant for the provision to be effective (section 29 Agricultural Tenancies Act 1995). Moreover, the tenancy agreement must not provide for the third party to be appointed without the consent or concurrence of both the landlord and the tenant. When a dispute has arisen, there must be a joint reference to the third party or, where one party makes the reference, notice must be served on the other informing him in writing that such a reference has been made. The person on whom the notice is served may, within four weeks from the notice, evoke the arbitration provisions of section 28 of the Agricultural Tenancies Act 1995. Under the provisions of that section, he must serve a notice in writing on the other party (i.e. in this situation the person who has given him the notice of reference made to the third party) specifying the dispute and stating that unless an arbitrator is appointed by agreement before the end of two months, he intends to apply to the President of the Royal Institution of Chartered Surveyors for the appointment of an arbitrator. If he fails to do so, then the independent third party named in the contract will have jurisdiction to settle the dispute.

2.6.7 Resolution by the courts

The jurisdiction of the courts is preserved except for cases where arbitration is mandatory. The landlord or the tenant may decide to institute court proceedings. The other party then has the option of applying to the court for a stay of proceedings which, if granted, will mean that the dispute will have to be resolved by arbitration (section 28(4) Agricultural Tenancies Act 1995).

2.7 Notices

2.7.1 Method of service

A notice or other document is duly given if it is delivered to the person or left at his proper address. The proper address is his last known address. Where the service is on a company, the proper address is the registered or principal office of the company (section 36 Agricultural Tenancies Act 1995).

Service may be effected by other means where there has been a prior written agreement between the person giving the notice and the recipient of the notice authorising service in that way. This might, for instance, include the giving of notice by facsimile or other electronic means. Unless there is such written agreement, notice transmitted by these means is not duly given.

2.7.2 Service on agents and companies

A notice or other document is duly given to a landlord if it is given to an agent or servant of his who is responsible for the control and management of the holding. Likewise, a notice or document is duly given to a tenant if it is given to his agent or servant responsible for the carrying on of a business on the holding. Service on an employee by either the landlord or tenant, who does not have these responsibilities is not sufficient. Where a notice or document is given to a company it must be served on the secretary or clerk of the company (section 36 Agricultural Tenancies Act 1995).

2.7.3 Change of landlord

The tenant must be informed of any change of landlord. Unless he has received notice that the former landlord has ceased to be entitled to receive the rents and profits and has been given the name and address of the new landlord any notice or document given to the old landlord will be deemed to have been properly served (section 36(7) Agricultural Tenancies Act 1995).

2.7.4 Notice under Landlord and Tenant Act 1987

Under section 48 of the Landlord and Tenant Act 1987, a landlord must

> 'by notice furnish the tenant with an address in England and Wales at which notices (including notices in proceedings) may be served on him by the tenant.'

Failure to do so will mean that rent or service charge due from the farm business tenant of the landlord will not be due until such notice is served.

Chapter 3
Agricultural Holdings Act Tenancies

3.1 Introduction

The law of agricultural holdings is extremely technical and complex. In particular, attention must be paid to the prescribed forms and time limits. This chapter can only give a superficial outline of the legislation.

3.2 Definitions

The current legislation protecting agricultural tenants whose tenancies pre-date 1 September 1995 is the Agricultural Holdings Act 1986. The legislation provides security of tenure for the tenant, regulates the terms of the tenancy and provides for compensation for the tenant on its termination. Where a tenancy was granted before 12 July 1984 there are succession provisions.

An agricultural holding is a letting of agricultural land which is used for the trade or business.

Agriculture

'includes horticulture, fruit growing, seed growing, dairy farming and livestock breeding and keeping, the use of land as grazing land, market gardens and nursery grounds, and the use of land for woodlands where the use is ancillary to the farming of land for other agricultural purposes.'
(section 96 Agricultural Holdings Act 1986)

Livestock

'includes any creature kept for the production of food, wool, skins, or fur or for the purpose of its use in the farming of land as the carrying on in relation to land of any agricultural activity.'
(section 96 Agricultural Holdings Act 1986)

Grazing of other animals (e.g. horses) by way of business is using land for agriculture (*Rutherford* v. *Maurer* [1961] 2 All ER 755). (The grazing was of horses in connection with a riding school not comprised in the lease.)

Buildings can amount to an agricultural holding.

Where land is let for both agricultural and non-agricultural uses it will be either wholly an agricultural holding or not at all. If the actual or contemplated use of the holding at the time of the contract and subsequently is predominantly agricultural it will be an agricultural holding (section 1(2) Agricultural Holdings Act 1986; *Howkins* v. *Jardine* [1951]

1 All ER 320). A non-agricultural tenancy will not become agricultural by a change of use contrary to the contract, unless the landlord agreed to, or acquiesced in, the change (section 1(3) Agricultural Holdings Act 1986).

Subtenancies are not protected as against the head landlord, although there may be protection against the mesne landlord. If the head tenancy terminates by notice to quit, given either by the landlord, or tenant the subtenancy will automatically terminate (*Pennell* v. *Payne* [1995] 2 All ER 592). However a surrender of the mesne tenancy will not determine the subtenancy.

3.3 The tenancy agreement

Many agricultural tenancies are oral. If there is no written agreement, or if there is one but it does not contain one or more of a list of items set out in the First Schedule of the 1986 Act, either party may refer the terms of the tenancy to arbitration (section 6 Agricultural Holdings Act 1986). The matters set out in the First Schedule are

(1) The names of the parties
(2) Particulars of the holding
(3) Any agreed terms
(4) The rent reserved and dates on which it is payable
(5) Liability for rates
(6) A covenant to return the full equivalent manurial value of any crops destroyed by fire
(7) A covenant by the tenant to insure against damage by fire or dead stock on the holding and all harvested crops grown on the holding for consumption
(8) A forfeiture clause
(9) A covenant by a tenant not to assign sub-let or part with possession of the holding or any part of it without the landlord's consent in writing

Before demanding an arbitration, the other party must be requested to enter into a written agreement containing the items in the First Schedule. Once such a request is made the tenant may not without the landlord's consent in writing assign, sub-let or part with possession of the holding or any part of it during the period while the determination of the terms of the tenancy is pending.

3.4 Fixed equipment

3.4.1 Maintenance, repair and insurance of fixed equipment

Incorporated into every tenancy agreement, except in and so far as the written tenancy agreement provides to the contrary, are provisions allocating the responsibility of the maintenance repair and insurance of fixed equipment between the parties (section 7 Agricultural Holdings Act 1986; Agriculture (Maintenance, Repair and Insurance of Fixed Equipment) Regulations 1973 (SI 1973/1473 amended by SI 1988/281). In broad terms the landlord is responsible for repairs and replacements to main structures of buildings, or to mains, sewage disposal systems and reservoirs and to insure the buildings against loss or damage by fire.

The tenant's obligations are 'except insofar as such liabilities fall to be undertaken by the landlord . . . to repair and to keep and leave clean and in tenantable repair, order and condition' most other items of fixed equipment on the holding including such things as hedges, ditches, roads and ponds.

The regulations allow a party, on the failure of the other to do his repairs, to carry them out himself and to recover the reasonable costs from the other, provided one month has elapsed after sending a written request to the other to do the repairs. The tenant's right to undertake replacements on the landlord's default is subject to monetary limits.

Where a written tenancy agreement departs substantially from the model clauses, it is open to either the landlord or tenant to go to arbitration seeking to bring the tenancy in line with the regulations. The arbitrator has a discretion whether to vary the tenancy. If he does, he may also vary the rent (section 8 Agricultural Holdings Act 1986).

3.4.2 Provision of fixed equipment

The tenant may apply to the Agricultural Land Tribunal for a direction that the landlord provides, alters or repairs fixed equipment so as to enable the tenant to comply with any statutory obligations (section 11 Agricultural Holdings Act 1986). This is an important provision when tenants are faced with expense in complying the pollution regulations or animal welfare provisions.

3.4.3 Removal of fixed equipment

Certain fixtures and buildings installed by the tenant on the holding for which he is not entitled to compensation, remain his property and may be removed by him, provided they were not installed under a tenancy obligation. Before removing them the tenant must have paid his rent, fulfilled his other obligations under the tenancy agreement and given the landlord the opportunity to purchase them. In removing the fixture the tenant must not do any avoidable damage and any damage he does do must be made good. Fixtures must be removed within two months of the termination of the tenancy (section 10 Agricultural Holdings Act 1986).

3.5 Rent

3.5.1 *Rent review*
The rent at the beginning of the tenancy is a matter for agreement between the parties. The 1986 Act provides a mechanism for rent reviews to take place every three years. The landlord or the tenant may demand an arbitration to fix the 'rent properly payable' for the holding, provided the rent change sought will not take effect during the first three years of the tenancy, or before three years have elapsed since the rent was last changed, or directed by an arbitrator to stay unchanged. Usually no arbitrator is appointed because the parties agree a new rent, but any party seeking a rent review is advised to serve an arbitration demand as a safeguard. The demand to the arbitrator must require the rent payable to be referred to arbitration 'as from the next termination date' (section 12 Agricultural Holdings Act 1986). This is defined by reference to the earliest date at which the tenancy could have been terminated by notice to quit. An example may make this clearer. If a tenancy runs from 11 October, in order to increase the rent on 11 October 2003, the notice must be served before 11 October 2002.

3.5.2 *Rent formula*
The 1986 Act sets out a rent formula of amazing complexity which it is almost impossible to apply (Schedule 2 Agricultural Holdings Act 1986). The arbitrator must

(1) Fix 'the rent at which the holding might reasonably be expected to be let by a prudent and willing landlord to a prudent and willing tenant'
(2) Taking into account
 (a) all relevant factors
 (b) the terms of the tenancy
 (c) the character, situation and locality of the holding
 (d) the productive capacity and related earning capacity of the holding with the fixed equipment and other facilities on it, assuming 'a competent tenant practising a system of farming suitable to the holding' and
 (e) rents of comparable holdings let on similar terms, including tendered rents likely to become payable
(3) Disregarding
 (a) evidence about comparables (if any), any element in the rent due to 'appreciable scarcity' of such holding available for letting or any element due to the tenant having the convenience of other land close by, or any allowance or reduction made for charging a premium
 (b) certain tenant's improvements high farming, grant-aided landlord's improvements, the fact that the tenant is a sitting tenant, and any disrepair or deterioration of the holding

Marriage value can be taken into account on the subject holding (*Childers v. Anker* [1995] EGCS 116.

3.5.3 Appointment of an arbitrator

The parties can chose an arbitrator by agreement, or either party can ask the President of the Royal Institution of Chartered Surveyors to appoint one from a panel set up by the Lord Chancellor. It is important for the party seeking the rent arbitration that before the 'termination date' either a new rent should be agreed, or an arbitrator appointed by agreement, or an application made to the RICS before that date to appoint an arbitrator. If not, the arbitration demand lapses (section 12 Agricultural Holdings Act 1986).

3.5.4 Rent increase for landlord's improvements

Where the landlord carries out improvements on the holding 'at the request of, or in agreement with, the tenant' or under certain statutory provisions, he is entitled to increase the rent, regardless of the three year rule mentioned above, provided he gives the tenant written notice within six months from completion of the improvement (section 13 Agricultural Holdings Act 1986). The increase is to be 'by an amount equal to the increase in the rental value of the holding attributable to the carrying out of the improvement' but if the landlord gets grant aid the rent increase is reduced proportionately. Any dispute is settled by arbitration.

3.5.5 Cultivation of the land

Underpinning the agricultural holdings legislation is the concept of good husbandry. However, there is no express obligation in the Act to comply with the rules of good husbandry. On the other hand, if the landlord obtains a certificate of bad husbandry from the Agricultural Lands Tribunal, he is entitled to serve a incontestable notice to quit.

There are two statutory provisions relating to the farming of the land. Section 14 allows either landlord or tenant to go to arbitration to seek to have the tenancy agreement modified as regards the amount of land to be maintained as permanent pasture in order to secure the full and efficient farming of the holding. Section 15 provides that, notwithstanding any agreement to the contrary, the tenant may dispose of the produce of the holding, other than manure, provided he makes suitable and adequate provision to return to the holding the full equivalent manurial value and to practise any system he likes of cropping the arable land, provided he protects the holding from injury or deterioration.

An important qualification is that these rights do not apply in the last year of the tenancy or in any period after the tenant has given or received notice to quit. In that period he must not sell or remove from the holding any manure, compost, hay or straw or any roots grown for consumption on the holding, without the landlord's consent.

3.6 Security of tenure

3.6.1 Statutory restrictions on notices to quit
Security of tenure is achieved by treating most tenancies of agricultural holdings as yearly tenancies and then imposing statutory restrictions on the operation of notices to quit.

Section 2 of the 1986 Act converts agreements for one year, or less than a year, into yearly tenancies even if they purport to be merely licences. Licences which are purely gratuitous or do not confer exclusive possession will not however be converted. A tenancy for a fixed term of two or more years does not end automatically at the end of the fixed term, but continues as a tenancy from year to year (section 3 Agricultural Holdings Act 1986).

Yearly tenancies, whether originally so granted or converted into yearly tenancies by section 2 or section 3 will continue unless a notice to quit is served of between 12 and 24 months expiring at the completed year of the tenancy (section 25 Agricultural Holdings Act 1986). In most cases the tenant is then entitled to serve a counter notice (section 26 Agricultural Holdings Act 1986).

The statutory conversion does not apply if the lease was granted after 12 September 1984 and the tenant dies during the currency of the fixed term. The lease will expire automatically on the term date specified in the lease, unless the tenant dies during the last year of the lease, when the lease will continue for a further year (section 4 Agricultural Holdings Act 1986).

It is not possible to contract out of the statutory provisions. Any attempt to do so will be invalid as contrary to public policy (*Johnson* v. *Moreton* [1978] 3 All ER 37).

3.6.2 Length of notice
As stated above, the general rule is that at least 12 months notice to quit must be given expiring, at a term date. The main exceptions to this are

(1) The tenant's insolvency
(2) The tenancy agreement validly contains a clause for resumption of the whole or part of the holding for a non-agricultural purpose
(3) Notice to a subtenant
(4) The tribunal or arbitrator specifies a date at the end of the tenancy upon issuing a certificate of bad husbandry (section 25(4) Agricultural Holdings Act 1986), or upon the tenant failing to comply with a notice to do work (Agricultural Holdings (Arbitration on Notices) Order 1987 SI 1987/710)
(5) A tenant can serve a short notice (at least six months) after a rent increase award (section 25(3) Agricultural Holdings Act 1986)

3.6.3 Counter notices
The tenant's main protection is his right to serve a counter notice to a

Agricultural Holdings Act Tenancies

notice to quit. This notice must be served within one month of the service of the notice to quit. The effect is that the notice to quit will not operate unless the landlord obtains the consent of the Agricultural Lands Tribunal (section 26 Agricultural Holdings Act 1986).

The Tribunal may give consent to the operation of the landlord's notice to quit if he applies within one month of the counter notice and proves one or more of the statutory grounds for possession (section 27 Agricultural Holdings Act 1986). The grounds are that he requires possession

(1) in the interests of good husbandry; or
(2) sound estate management; or
(3) to carry out agricultural research, education etc.; or
(4) for statutory allotments; or
(5) because greater hardship would be caused by withholding than by giving consent to the notice to quit; or
(6) for certain non-agricultural uses of the holding which do not require planning permission.

Even if the landlord is able to prove a ground for possession he may not obtain consent because the Tribunal is required to withhold consent 'if in all the circumstances it appears to them that a fair and reasonable landlord would not insist on possession'.

3.6.4 Incontestable notices

There are eight cases where the tenant has no right to serve a counter notice. All the tenant can do is to challenge by arbitration whether the ground stated in the notice to quit in fact exists in relation to Cases B, D or E (Agricultural Holdings (Arbitration on Notices) Order 1987, SI 1987/710). He has one month in which to demand arbitration. The arbitrator must be appointed within three months thereafter.

Incontestable notices can be served under the following cases.

(1) *Case A – retirement from small holding:* a tenant has attained age 65, suitable alternative accommodation is available and the tenancy agreement is made expressly subject to this case.
(2) *Case B – planning permission:* the land is required for a non-agricultural use for which planning permission has been obtained or is not needed by virtue of an enactment other than the Town and Country Planning legislation.
(3) *Case C – bad husbandry certificate:* a certificate of bad husbandry was issued by the Agricultural Land Tribunal and notice to quit is served within six months of that certificate.
(4) *Case D – notices to remedy breaches of tenancy:* on the grounds that either
    (a) the tenant has failed to comply with a notice requiring him to pay overdue rent within two months; or

(b) the tenant has failed to comply within a reasonable time with a notice to remedy a breach of tenancy capable of being remedied.
(5) *Case E – irremediable breach of tenancy:* the landlord is materially prejudiced by breach of the tenancy agreement which is not capable of being remedied.
(6) *Case F – insolvency of the tenant.*
(7) *Case G – death of the tenant:* the death of the sole or surviving tenant is incontestable provided the notice to quit is served within three months of receiving written notification of the death, or an application for a succession tenancy. The notice to quit will end the tenancy but a close relative might be granted a succession to the tenancy.
(8) *Case H – Minister's amalgamation scheme:* the Secretary of State for Rural Affairs or Secretary of State for Wales requires the land for an amalgamation scheme. The tenancy agreement must acknowledge that the property might be required for this purpose (Schedule 3 Agricultural Holdings Act 1986).

Cases A to G require the notice to quit to state the statutory reason for possession.

3.6.5 Notice to pay rent

In order to serve such a notice, not only must the rent be overdue but a notice must have been served requiring the tenant to pay the rent within two months. If the tenant fails to pay the rent within two months, then the ground under Case D is established and there is no relief available. Subsequent payment by the tenant will not remove the landlord's right to possession, even if the rent is paid before the notice to quit is served.

It is important that the landlord has given the tenant an address in England or Wales at which notices, including notices in proceedings, can be served on the landlord by the tenant (section 48 Landlord and Tenant Act 1987). If this is not done, any rent due is not treated as due and so a Case D Notice cannot be validly served.

Any written demand for rent must contain the landlord's address (section 47 Landlord and Tenant Act 1987).

3.6.6 Notices to remedy other breaches of tenancy

The rules concerning notices to remedy (other than those relating to rent) are extremely complex. Notices to remedy must be in the prescribed form (paragraph 10 Schedule 3 Agricultural Holdings Act 1986). The notice must give the tenant a reasonable time to comply with all the items in the notice (*Wykes v. Davis* [1975] 1 All ER 399).

Where the notice is a notice to do work, there are additional rules as listed below.

- A valid notice cannot be served within 12 months of a previous notice to do work unless the previous notice was withdrawn with the tenant's

written agreement. At least six months have to be given for doing any item of work.
- ❐ The tenant may go to arbitration on any question arising under the notice, such as his liability to do the work specified. He may also invite an arbitrator to exercise his powers to delete 'any unnecessary or unjustified' items in the notice, or to substitute 'a different method or material' for any specified in the notice.
- ❐ Even if the notice is modified by the arbitrator, it does not modify the tenancy obligations of the tenant. The landlord can pursue any other remedies open to him and the tenant may have to meet a dilapidations claim at the end of the tenancy if he does not do the work as required by the tenancy agreement.

If a tenant receives a notice to quit under Case D for failure to comply with the notice to do work, the tenant can serve a counter notice within one month. The notice to quit will then not take effect unless the Agricultural Land Tribunal consents to its operation. The Tribunal has to decide whether a fair and reasonable landlord would insist on possession. If after the counter notice the tenant goes to arbitration to test the reason given in the notice to quit, the counter notice is nullified, but should the Arbitrator uphold the notice to quit the tenant can give a further counter notice within one month from the delivery of the award (section 28 Agricultural Holdings Act 1986).

3.7 Compensation

3.7.1 Adherence to time limits and procedures
The 1986 Agricultural Holdings Act lays down strict time limits for making claims for compensation. The correct procedures must be followed. Disputed claims are settled by arbitration (section 83 Agricultural Holdings Act 1986).

3.7.2 Improvements by the tenant
Schedule 7 of the Agricultural Holdings Act 1986 divides long-term improvements begun on or after 1 March 1948 into two categories. Those included in Part I are improvements for which the consent of the landlord is required before compensation is payable. Those in Part II require the consent of the landlord or approval of the Agricultural Land Tribunal. Short-term improvements are governed by Schedule 8. For these improvements the tenant is entitled to compensation whether or not he obtains consent.

3.7.3 Landlord's consent required
Improvements in Part I are of a specialised nature. They are the making or planting of osier beds, water meadows, watercress beds, planting of hops,

orchards or fruit bushes, warping or weiring of land, making of gardens and provision of underground tanks. Before the tenant is entitled to compensation, the landlord must give his consent. He can attach any conditions to that consent.

The measure of compensation is

> 'an amount equal to the increase attributable to the improvement and the value of the agricultural holding as a holding, having regard to the character and situation of the holding and the average requirements of tenants reasonably skilled in husbandry'.
>
> (section 66(1) Agricultural Holdings Act 1986)

It is quite common for the parties to agree a value at the time of the improvement and to depreciate it on a percentage basis over an agreed number of years.

3.7.4 Consent of landlord or Tribunal required

For improvements listed in Part II of Schedule 7 the tenant will require the landlord's consent if he is to be entitled to compensation, or, if the consent is refused or is given with conditions the tenant does not like, the tenant may apply to the Agricultural Lands Tribunal for its approval. Should the tribunal give its approval, the landlord may elect to carry out the improvement himself. If he does not do so, the approval will count as the landlord's consent. The tribunal may attach conditions to its approval. The rules and procedures are set out in section 68 Agricultural Holdings Act 1986.

The measure of compensation is the same as for improvements under Part I of Schedule 7.

3.7.5 No consent needed

The tenant will be entitled to compensation for any improvements he does that are listed in Schedule 8 of the 1986 Act. For the most part these are acts of good husbandry and the tenant is compensated for leaving behind some benefit from them at the end of the tenancy.

Section 66(2) provides that the amount of compensation for short-term improvements is the value of the improvement to an incoming tenant calculated in accordance with the prescribed regulations currently in force. The regulations currently in force are the Agriculture (Calculation of Value for Compensation) Regulations 1978 (SI 1978/809) as amended by SI 1980/751, the Agriculture (Calculation of Value for Compensation) (Amendment) Regulations 1981 (SI 1981/822) and the Agriculture (Calculation of Value for Compensation) (Amendment) Regulations 1983 (SI 1983/1475). Part I deals with short-term improvements and Part II with tenant right matters which are set out in Part II of Schedule 8.

3.7.6 Disturbance compensation

In addition to any compensation a tenant may be entitled to at the end of the tenancy for improvements and tenant right matters he will be entitled to disturbance compensation unless the tenancy was surrendered or the tenancy ended with a notice to quit under Cases C, D, E, F or G as set out in section 3.6.4.

Disturbance compensation is usually equal to one year's rent. Where extra expense on quitting is incurred it may be up to two years' rent. Additional compensation of a further four years' rent may be claimed where the landlord is recovering possession for non-agricultural purposes. If no ground is stated the landlord will be liable to pay this additional compensation no matter what the actual ground is for possession (sections 60 and 61 Agricultural Holdings Act 1986).

3.7.7 High farming
Section 70 of the Agricultural Holdings Act 1986 allows claims for high farming but these provisions are seldom relied on in practice.

3.7.8 Game damage
The tenant of an agricultural holding can claim compensation from his landlord if his crops are damaged by any wild animals or birds that the landlord (or anyone claiming under him) has the right to kill, provided that the tenant has not permission in writing to kill them (section 20 Agricultural Holdings Act 1986). The compensation can only be claimed if notice in writing is given to the landlord within one month after the tenant first became, or ought reasonably to have become, aware of the damage and a reasonable opportunity is given to the landlord to inspect the damage before the crop is reaped (or if the damage is to a gathered crop, before it is removed). Written particulars of the claim must also be given within one month after the expiry of the year in respect of which the claim is made. The year ends on 29 September unless otherwise agreed.

In cases where the landlord does not have the sporting rights, he is entitled to be indemnified against claims for game damage by whoever has them.

3.7.9 Landlord's compensation
The landlord is entitled to compensation from the tenant for any dilapidation or deterioration of, or damage to, any part of the holding or anything in or on the holding caused by non-fulfillment by the tenant of his responsibilities to farm in accordance with the rules of good husbandry. The amount is the cost of making good as at the date of quitting but must not exceed the diminution in the value of the landlord's reversion (section 71 Agricultural Holdings Act 1986). If the neglect is so bad that the general value of the holding has been reduced, the landlord may claim for the decrease in value (section 72 Agricultural Holdings Act 1986).

3.8 Succession tenancies

3.8.1 Tenancies capable of succession

The laws enabling close relatives to obtain succession tenancies on the death or retirement of tenants are set out in Part IV of the 1986 Act in Schedule 2. They apply

- to tenancies granted before 12 July 1984;
- to any tenancy granted after that date as a first statutory succession;
- any tenancy granted after that date and before 1 September 1995 where the parties have contracted in writing into the statutory provisions; and
- any tenancy granted after that date to a person who immediately before was a tenant of the holding which comprised the whole or a substantial part of the land within the new tenancy.

There is no automatic right of succession. When a sole or sole surviving tenant of an agricultural holding dies, certain close relatives or treated children (see the definition in section 3.8.2) of the deceased may apply within three months of the death to the Agricultural Lands Tribunal for a direction entitling them to a new tenancy of the deceased's holding. One such near relative can be nominated to apply for a succession tenancy during the tenant's lifetime, once the tenant reaches 65. Before an applicant can succeed, a tribunal must be satisfied that he or she is both eligible and suitable for succession and they must also consider any case put forward by the landlord for recovery of possession. There can only be two successions to a holding under the Act.

3.8.2 Succession on death

There will be no claim for succession in the following circumstances

(1) The tenancy post dates 11 July 1984
(2) There have already been two successions under the Act (section 37)
(3) The tenancy was subject to a valid notice to quit which could not be effectively challenged
(4) The deceased's tenancy was a fixed term with more than 27 months still to run (section 36(2))
(5) The holding is a statutory smallholding or one under certain ex-servicemen's charitable trusts (section 38(4) and (5))

If the exceptions do not apply then an eligible person may apply for a succession tenancy.

In order to be eligible the applicant must satisfy three tests as listed below.

(1) He must be within the qualifying degrees of kindred to the deceased tenant namely – spouse, brother, sister, child or treated child. A treated child is a person 'who in the case of any marriage to which the

deceased was at any time a party, was treated by the deceased as a child of the family in relation to that marriage' (section 35 Agricultural Holdings Act 1986).

(2) The applicant must for at least five out of the last seven years of the deceased's life have derived his only or principal source of livelihood from his agricultural work on the holding or on an agricultural unit of which the holding forms a part. Up to three years on a full time course at an establishment of further education in any subject will count towards the qualifying five years (paragraph 2 Schedule 6 Agricultural Holdings Act 1986). In the case of the deceased's widow the agricultural work of the deceased will count as hers (section 36(4) Agricultural Holdings Act 1986).

The livelihood rule is not an absolute rule. Section 42 provides that if the rule is not fully satisfied, but is satisfied to a material extent, the tribunal may treat the applicant as qualifying under the rule if it considers it fair and reasonable to do so.

(3) The applicant must not be the occupier of a commercial unit of agricultural land. Occupation under a farm business tenancy of less than five years is disregarded (paragraph 6(12)(d) Schedule 6).

Commercial unit is defined to mean 'a unit of agricultural land which is capable, when farmed under competent management, of producing a net annual income of an amount not less than the aggregate of the average annual earnings of two full time, male agricultural workers aged 20 or over' (paragraph 3 Schedule 6 Agricultural Holdings Act 1986). Whether a unit is commercial may be a difficult question. The Secretary of State for DEFRA periodically publishes average agricultural earnings and also makes Unit of Production Orders annually to enable the net annual income for various kinds of farming enterprises to be assessed (paragraph 4 Schedule 6 Agricultural Holdings Act 1986).

Schedule 6 also specifies rules for determining whether a joint occupier (for example a partner) has a sufficient legal interest or share in a commercial unit to be disqualified for succession and counts occupation by a spouse, or by a company controlled by the applicant or spouse, as occupation by the applicant. An applicant may not succeed to more than one commercial unit.

The application to the tribunal for succession must be made within the period of three months beginning with the day after the date of death on the prescribed form (section 39 Agricultural Holdings Act 1986).

Even where the tribunal is satisfied that an applicant is eligible, it must decide whether he is also a suitable person to become the tenant of the holding. If there is more than one eligible applicant, then (unless one has been designated in the will or the landlord agrees to a joint tenancy) the tribunal must decide which is the more or most suitable (section 39(6) Agricultural Holdings Act 1986).

In deciding on suitability, the tribunal must have regard to all relevant matters including the extent of the applicant's training and experience in agriculture, his age, health and financial standing and also the views (if any) stated by the landlord on his suitability.

If the landlord wants to resist the succession tenancy he should serve a notice to quit under Case G (see section 3.6.4) within three months of being given written notice of the death. If an application is made for a succession tenancy the landlord is entitled to reply on the prescribed form disputing the applicant's eligibility or suitability. He may also apply for consent to the operation of his notice to quit. Consent may be given on any of the grounds in section 27 of the Act, but even if he can prove a ground, the tribunal may still refuse possession because of the fair and reasonable landlord test. If consent is given to the notice to quit, or if the tribunal determines the applicant is not eligible or suitable, the succession application will be dismissed (section 44 Agricultural Holdings Act 1986). If there is an eligible and suitable applicant and consent is not given to the notice to quit, the tribunal must direct a succession to the tenancy in favour of the applicant.

As only two successions are allowed under the Act, it is important for landowners that any succession which does take place counts as one under the Act. A voluntary grant to an eligible person who has not applied for succession will not count. The close relative should apply under the Act. If after the time limits for applying for succession have elapsed he is, or has become, the sole remaining applicant, a voluntary grant of a tenancy can then take place under the Act without a hearing (section 37(1)(b) Agricultural Holdings Act 1986). In these situations, it is prudent to record the succession under the Act by a recital in the tenancy agreement.

A tenancy directed by the tribunal to a successful applicant will be a new tenancy on the same terms as the deceased's at the end of his tenancy (section 47 Agricultural Holdings Act 1986). However, whether or not the deceased had a covenant against assignment, subletting or parting with possession without the landlord's consent, the new tenancy will have such a covenant included by law. The new tenancy will start from the date on which the notice to quit ends the deceased's tenancy or the equivalent date if no notice was served (sections 45 and 46 Agricultural Holdings Act 1986).

Either the landlord or the successful applicant may demand an arbitration to review the terms of the tenancy, including the rent. The arbitrator may make any justifiable variations and in any event must include in the tenancy provision for the payment of the usual ingoings to the landlord and payment to the new tenant of any compensation for dilapidations recovered by the landlord at the end of the deceased's tenancy in respect of any items the new tenant is required to make good. If the tenancy terms are varied, the arbitrator may adjust the rent accordingly (section 48(6) Agricultural Holdings Act 1986).

A rent arbitration must be demanded irrespective of the time of the last previous rent review. The Arbitrator must fix the rent properly payable in accordance with the formula in Schedule 2 of the Agricultural Holdings Act 1986.

3.8.3 Succession on retirement of tenant

A succession counting as one of the two permitted under the Act may take place by agreement during a tenant's lifetime. The tribunal is not involved if there is an agreement between the landlord, the tenant and the relative of the tenant who is within the qualifying degrees of kindred, that the tenant retires from his tenancy and a new tenancy is granted to the relative or assigned to him or her (section 37 Agricultural Holdings Act 1986).

In the absence of an agreement a tenant of a tenancy which started before 12 July 1984 or of a first succession tenancy may, on or after reaching 65, or earlier in the case of bodily or mental incapacity, serve on the landlord a retirement notice nominating one eligible close relative (section 49 Agricultural Holdings Act 1986). The nominated close relative may then apply to the tribunal, within one month, for a succession tenancy (section 50 Agricultural Holdings Act 1986). The rules for eligibility and suitability are the same as for applications on the death of the tenant. If the tribunal grants succession, the tenant will retire from his tenancy and the succession tenancy will begin during his lifetime.

The tenant may serve only one retirement notice and nominate only one close relative (sections 51(2) and 49(1)(b) Agricultural Holdings Act 1986). Nobody but the nominated person may apply for succession and a prescribed form must be used. A tribunal has to consider whether greater hardship would be caused by giving the direction than by refusing the nominated successor's application (section 53(8) Agricultural Holdings Act 1986). If the applicant fails the tenancy continues and the applicant cannot apply again on the tenant's death (section 57(4) Agricultural Holdings Act 1986).

A notice to quit which is effective (e.g. because it is an incontestable notice or the tribunal has given consent) cannot be defeated by the subsequent service of a retirement notice. Once a retirement notice has been served, an incontestable notice to quit may be validly served, provided it is given before the tribunal hearing begins and the incontestable ground existed before the retirement notice. Notice to quit based on Cases C and D will only prevail if the certificate of bad husbandry, or the notice to remedy or pay rent was applied for, or given, before the retirement notice (sections 51(4),(5) and 52 Agricultural Holdings Act 1986).

3.9 Disputes procedure

3.9.1 The right forum

Disputes under the Agricultural Holdings Act are resolved either by arbitration or the Agricultural Lands tribunal or the courts. It is important that the right forum is selected, otherwise the claim will fail.

3.9.2 Arbitration

Arbitration under the 1986 Act is a separate statutory code and is substantially different from arbitration under what is now the Arbitration Act 1996.

The following disputes are compulsorily referable to arbitration under section 84 and Schedule 11 of the Agricultural Holdings Act 1986.

(1) Disputes as to whether an agreement has been converted into a yearly tenancy under section 2
(2) Securing a written tenancy agreement under section 6
(3) Varying or modifying repairing obligations under sections 7, 8 and 9
(4) Revising the rent under sections 12 and 13
(5) Questions relating to game damage under section 20
(6) Notice to remedy and Notices to Quit under Cases A, B, D or E of Schedule 3
(7) Claims arising on or out of the termination of the tenancy (section 83)

3.9.3 Agricultural Lands Tribunal

The Agricultural Lands Tribunal has jurisdiction for

(1) The giving or withholding of consent to the operation of a Notice to Quit
(2) Granting a Certificate of Bad Husbandry
(3) Applications for succession tenancies on death or retirement
(4) Determining whether a holding is a market garden
(5) Approval for long-term improvements where the landlord refuses his consent
(6) Directions to the landlord to carry out improvements
(7) Directions to a landlord to provide or repair fixed equipment

3.9.4 The courts

The inherent jurisdiction of the courts is preserved except where the agricultural holdings legislation provides to the contrary. An application may be made through the case stated procedure to the Divisional Court from the Agricultural Lands Tribunal and to the county court from an arbitrator.

Chapter 4
Residential Protection of Farm Workers

4.1 Introduction

The residential protection granted to farm workers is extremely complex. Amongst other matters it depends on when the tenancy was granted and how much rent is payable under the agreement.

Where a farm worker has a tenancy, as distinct from a service occupancy, and pays sufficient rent, he will be protected as other residential tenants by the Rent Act 1977, if the tenancy was granted before 15 January 1989. If granted after that date he will be protected by the Housing Act 1988 as an assured agricultural occupant even if he pays a low rent.

If, on the other hand he has merely a service occupancy or his rent is below the threshold for a Rent Act or Housing Act tenancy, he will be protected either by the Rent (Agriculture) Act 1976 or the assured agricultural occupancy provisions of the Housing Act 1988.

4.2 The Rent (Agriculture) Act 1976

4.2.1 Application of the Act
This Act continues to apply to most tenancies created before the commencement of the Housing Act 1988. It is retrospective and applies to tenancies created before the commencement of the 1976 Act. A qualifying worker with a relevant licence or tenancy of a dwelling which is, or has been, in qualifying ownership during the subsistence of the tenancy will be a protected occupier in his own right. On termination of his contractual tenancy or licence he will be a statutory tenant in his own right.

4.2.2 Qualifying worker
A person is a qualifying worker if he has worked whole time in agriculture at any time for not less than 91 out of the preceding 104 weeks (paragraph 1 Schedule 3 Rent (Agriculture Act) 1976). A week must be at least 35 hours. Certain weeks count towards the entitlement even if he does not work during them (any week where he is absent from work or works less than the 35 hours with the consent of his employer, holidays to which he is entitled and weeks in which he is absent due to injury or disease). If the worker is suffering from a qualifying injury or disease any week of absence will count as a week of whole time work in agriculture.

A qualifying injury or disease is an injury caused by an accident arising out of and in the course of agricultural employment or an injury prescribed in relation to agricultural workers under the social security legislation.

Such a worker will be protected even if he has not completed 91 weeks agricultural work (section 2(2) Rent (Agriculture) Act 1976).

If after 91 weeks the worker recovers, the time when he was incapacitated will count as qualifying weeks. However, should he recover before the 91 weeks he will lose the protection until the weeks during which he was incapacitated together with the working weeks amount to 91.

There are special rules for permit workers. These are employees who due to physical or mental incapacity are prevented from earning the minimum rate of pay prescribed by the Agricultural Wages Order. The Agricultural Wages Board may give a permit exempting such a worker from the statutory restrictions under the Agricultural Wages Act 1948, which apply to other agricultural workers. A permit worker does not have to work a 35 hour week to count as a qualifying worker.

The worker must be employed in agriculture under a contract of employment. The 1976 Act does not confer protection on independent agricultural contractors.

Agriculture is defined by the Rent (Agriculture) Act 1976 and the definition is different from that in the Agricultural Holding Act 1986.

Agriculture includes

> 'dairy farming, livestock keeping and breeding, the production of any consumable produce grown for sale or consumption, the use of land as grazing, meadow or pasture, orchards, osier beds, market gardens or nurseries and forestry'.
>
> (section 1 Rent (Agriculture) Act 1976)

As the word used in the definition is 'includes' the list is not exhaustive. Arable farming would be included but not activities which are not related to commercial farming purposes.

Livestock is defined to include any animal which is kept for the production of food, wool, skins or fur or for its use in the carrying on of any agricultural activities.

Unprotected workers include

- Gamekeepers
- Those employed in fish farming
- Stable lads
- Shop assistants in a farm shop

On the other hand, an employee involved in general farm work, e.g. repairing farm machinery will be protected.

4.2.3 *Qualifying ownership*

A dwelling is in qualifying ownership if the occupier is employed in agriculture, or has been during the period covered by the licence or tenancy and the occupier's employer is either the owner of the dwelling-house or has made arrangements with the owner for it to be used as housing accommodation for his agricultural employees (paragraph 3 Schedule 3 Rent (Agriculture) Act 1976).

4.2.4 Relevant licence or tenancy

The occupier must have exclusive occupation, either as a tenant or licensee of a dwellinghouse as a separate dwelling (paragraph 1 Schedule 2 Rent (Agriculture) Act 1976). The conditions necessary for a tenant to have protection under the Rent Act 1977 have to be fulfilled except

(1) A service occupancy is covered
(2) There may be no rent or the rent may be less than that needed to afford protection under the Rent Act 1977
(3) The dwelling may be comprised in the letting of an agricultural holding

4.2.5 Continuation of protection

A protected occupier in his own right will continue to be protected where (1) there is a surrender and regrant of a tenancy and (2) where a new licence or tenancy is granted in consideration of his giving up possession of another dwellinghouse of which he was a protected occupier.

More surprisingly the protection will continue where the occupier ceases to be employed in agriculture, retires or is dismissed, as long as he remains in the property.

An employer taking on a new worker may find that the employee will qualify at once as a protected agricultural occupant. The qualifying employment need not be with the same employer. If an agricultural worker has at any time worked for 91 out of the previous 104 weeks he achieves the status of a qualifying worker. This status attaches to him and if he is offered accommodation by a new employer of premises in qualifying ownership under a relevant tenancy or licence he will be protected.

4.2.6 Nature of the statutory tenancy

The protected occupancy will usually end on the termination of the worker's employment or in accordance with any contractual provisions or common law notice for termination. The tenant then become a statutory tenant in his own right (sections 2(4) 4(1) Rent (Agriculture) Act 1976).

This confers merely a personal right of occupation. It will therefore terminate if the tenant dies without anyone being entitled to a succession tenancy, or if he vacates the properly without a settled intention to return.

4.2.7 Rent

During the protected occupancy it is likely that no rent or a low rent will have been charged. Once the employment has determined, the employer may want to increase the rent.

Although the parties can agree a rent, it must not exceed either the registered fair rent or if there is no registered rent the weekly or other periodic equivalent of a rent of $1\frac{1}{2}$ times the rateable value (sections 11 and 12 Rent (Agriculture) Act 1976). Where the dwellinghouse had no rateable value on 31 March 1990 the rent must not exceed the rent at which the dwellinghouse might reasonably be expected to be let. Either party may apply for the registration of a fair rent. The Rent Acts (Maximum Fair Rent) Order 1999 limits the amount of any increase in a registered fair rent.

Where a fair rent has been registered the agreed rent can be raised to the level of the fair rent provided a notice of increase is served. If no rent has been registered and the agreed rent is less than the periodic equivalent of $1\frac{1}{2}$ times the rateable value of the dwelling the rent can be increased to that level after service of the requisite notice.

Agreed rent in excess of the upper limits imposed by the 1976 Act is not recoverable from the tenant.

No rent is payable under the statutory tenancy until there is an agreed rent or the notices of increase have been served.

In the absence of any agreement to the contrary rent is payable weekly in arrears (section 10 Rent (Agriculture) Act 1976).

4.2.8 Other terms

The underlying principle is that the terms of the original protected occupancy contract will continue to apply insofar as they are relevant to a statutory tenancy (section 10 and Schedule 5 Rent (Agriculture) Act 1976).

Where the original contract was a licence this is converted into a weekly tenancy and any implied terms of such a tenancy will apply.

Certain obligations are imposed on the landlord by the Act. He must continue to supply any services or facilities he previously supplied if these are reasonably necessary for the statutory tenant occupying the property. Section 11 of the Landlord and Tenant Act 1985 is applied. This imposes on the landlord an obligation to keep the structure and exterior of the property in repair and the installations for the supply of water, gas and electricity.

The tenant is under an obligation not to use the dwellinghouse or any part of it for any other purpose than a private dwellinghouse. Nor may the tenant assign, sublet or part with possession of the dwellinghouse or any part of it.

The tenant must allow the landlord reasonable access for the execution of repairs. He is also under an obligation to give at least four weeks notice to quit.

The parties may by an agreement in writing vary the terms of the statutory tenancy.

4.2.9 Security of tenure

Security of tenure is given to the tenant in much the same way as it is under the Rent Act 1977. In order to obtain possession the landlord must show that the contractual tenancy has been terminated and either a mandatory ground for possession exists (see section 4.2.10.1) or there is a discretionary ground (see section 4.2.10.2) and it is reasonable for the court to order possession (section 6 and Schedule 4 Rent (Agriculture) Act 1976).

Where a discretionary ground is relied on the court has power to stay or suspend an order for possession or to postpone the date of possession for as long as it thinks fit. The court may also adjoin the proceedings (section 7 Rent (Agriculture) Act 1976).

4.2.10.1 Mandatory grounds
These occur in the following circumstances.

(1) The person who granted the tenancy was an owner-occupier and the court is satisfied that the dwellinghouse is required as a residence for the original occupier or a member of his family who resided with him when he last occupied the dwelling as his residence. Notice must have been given at the start of the tenancy that possession might be recovered on this ground.
(2) The person who granted the tenancy acquired the dwelling for the purpose of occupying the residence on retirement and he now requires it as a retirement home or if he has died it is required by a member of his family residing with him when he died. Notice must have been given at the start of the tenancy that possession might be recovered on this ground.
(3) The occupier is guilty of the offence of overcrowding under the housing legislation.

4.2.10.2 Discretionary grounds
These are similar to the grounds under the Rent Act 1977. There are discretionary grounds for possession if

(1) Rent lawfully due is unpaid or the tenant is in breach of other covenants
(2) The tenant or his lodger or subtenant have been guilty of conduct which is a nuisance or annoyance to adjoining occupiers
(3) The condition of the dwellinghouse or furniture provided by the landlord has deteriorated due to the neglect or default of the tenant
(4) The tenant has without the consent of the landlord assigned, sublet or parted with possession of the dwellinghouse or any part of it
(5) The premises have been sublet at a rent exceeding the maximum recoverable under the 1976 Act or the Rent Act 1977
(6) The tenant has given notice to quit and the landlord has contracted to sell or let the dwellinghouse or taken steps which would be seriously prejudiced by being unable to obtain possession. (NB: a notice by the tenant to terminate his employment does not count as a notice to quit.)
(7) The dwelling is reasonably required by the landlord as a residence for

himself, his son or daughter, parents or grandparents. For this ground the landlord must not have become a landlord by purchasing the reversion of the dwellinghouse after 12 April 1976. The court must be satisfied that no greater hardship would be caused by granting possession than by refusing it.
(8) Suitable alternative accommodation is provided either by the landlord or the housing authority.

In the agricultural context the most important discretionary ground is that alternative accommodation is available for the tenant. The underlying principle of the Act is that agricultural efficiency should not be jeopardised by tying up agricultural accommodation when it is needed for new employees.

Where the alternative accommodation is provided by the landlord it must be suitable, taking into account the tenant's needs, as to the character and extent of the property and the tenancy must be on similar terms with the same security as his existing tenancy.

If the landlord is unable to provide alternative accommodation but requires vacant possession for another agricultural worker, who is needed in the interests of efficient agriculture, he can apply to the housing authority to rehouse the statutory tenant. Agricultural Dwellinghouse Advisory Committees (ADHACs) advise housing (i.e. local) authorities on the agricultural merits of the case. If the committee supports the landlord the housing authority must use its best endeavours to rehouse the displaced occupier.

The landlord will be entitled to possession if

(1) an offer of suitable accommodation is made, giving the tenant not less than 14 days in which to accept; or
(2) the housing authority have notified the tenant of an offer of alternative accommodation from a third party which the tenant has accepted within the time allowed; or
(3) the tenant acted unreasonably in refusing the housing authority's offer of alternative accommodation.

4.2.11 Succession rights

There is only one succession to a 1976 Act tenancy (section 4 Rent (Agriculture) Act 1976).

A spouse who was residing with the original tenant at his or her death becomes a statutory tenant as long as he or she occupies the dwelling as his or her residence. A cohabitant will be treated as if he or she were living with the original occupier as his or her wife or husband.

If the occupier had no spouse living with him in the dwelling at his death succession rights are given to any member of his family who was residing with him in the dwellinghouse at his death and had been doing so for the two preceding years. However, whilst a spouse obtains a statutory tenancy, since the commencement of the Housing Act 1988 the member of the family obtains the more limited protection of an assured tenancy (see section 4.3.1).

4.3 The Housing Act 1988

4.3.1 Application of the Act

The Housing Act 1988 created the assured tenancy. The assured agricultural occupancy is a variation of the assured tenancy. Licences or tenancies granted to farm workers after 15 January 1989 will normally be assured agricultural occupancies.

The Rent (Agriculture) Act 1976 continues to govern licences and tenancies granted before that date. A tenancy

(1) entered into pursuant to a contract before that date though granted afterwards;
(2) granted to a person who immediately before the grant was a protected or statutory occupier under the 1976 Act provided that the grant is made by the same landlord or licensor; and
(3) directed by the court to be a protected tenancy where an order for possession has been obtained on the basis of alternative accommodation

is also governed by the 1976 Act (section 34 Housing Act 1988).

4.3.2 The assured agricultural occupancy

In order to be an assured agricultural occupancy the licence or tenancy must satisfy the conditions for an assured tenancy or the only reason it does not is that the dwelling is granted at a low rent or the dwelling is included in an agricultural holding and is occupied by the person responsible for the farming of the holding. A licence, giving exclusive possession but being a service occupancy and therefore not an assured tenancy, if the other conditions are fulfilled, can be an assured agricultural occupancy (section 22 Housing Act 1988).

In addition, in order to be an assured agricultural occupancy the agricultural worker condition must be fulfilled. This involves complying with the landlord condition and the occupier condition under Schedule 3 of the Housing Act 1988 (see sections 4.3.2.1 and 4.3.2.2).

4.3.2.1 The landlord condition

The dwellinghouse must be, or have been, in qualifying ownership at some time during the subsistence of the tenancy or licence. Qualifying ownership means that the occupier must be employed in agriculture and his employer must be the owner of the dwellinghouse or have made arrangements with the owner to house the employer's agricultural workers.

4.3.2.2 The occupier condition

The occupier must be a qualifying worker or have been a qualifying worker at some time during the subsistence of the licence or tenancy. A qualifying worker is someone who has worked whole time in agriculture (or as a permit worker) for not less than 91 out of the preceding 104 weeks. Or the occupier must be incapable of full time or permit work in agriculture due to a qualifying injury or disease suffered in the course of his agricultural employment.

4.3.3 Succession rights

Succession rights under the Housing Act 1988 are achieved by deeming the qualifying widow or widower to fulfil the agricultural worker condition on the death of the qualifying worker. The spouse must have been residing with the previous occupier in the dwellinghouse immediately before his or her death. A spouse is deemed to include someone living with the deceased occupier as his or her wife or husband.

If there is no spouse, a member of the deceased's family will qualify if such member was residing in the dwellinghouse with the deceased at the time of and for two years before his death. A qualifying spouse or family member will obtain an assured agricultural occupancy.

4.3.4 Continuing protection

The agricultural worker condition will continue to be satisfied if a protected tenant accepts alternative accommodation in place of accommodation in which he was previously protected. He will also be protected if there is a surrender and regrant of a tenancy of the same dwelling.

4.3.5 Cessation of agricultural worker condition

The agricultural worker condition will continue to be fulfilled even if the worker takes up other employment, retires or is dismissed.

It will cease, however, if a fixed term tenancy of tied agricultural accommodation is assigned to someone who is not a qualifying agricultural worker. It will also cease if on the tenant's death the tenancy vested in a non-qualifying person. Unlike a tenancy protected under the Rent (Agriculture) Act 1976 the statutory tenancy is not a mere personal right but is a property right which can be transferred.

The 1988 Act implies an absolute covenant against assignment in all periodic tenancies of residential accommodation. If such property passes under the will or intestacy of the tenant, the landlord has a mandatory ground for possession provided he brings proceedings within 12 months of the tenant's death (Schedule 2 ground 7 Housing Act 1988).

The effect of the cessation of the agricultural worker conditions depends on whether the tenancy was an assured tenancy or merely deemed to be an assured tenancy (because the rent was too low or it was a licence etc.). If the latter, then the landlord, having given notice, will be able to obtain possession by court proceedings. However, if it is an assured tenancy he will have to prove a ground for possession. The most likely ground will be ground 16 (i.e. the dwellinghouse was let in consequence of the tenant's occupation which has now ceased).

4.3.6 Security of tenure
After the original fixed term tenancy is terminated the occupant is entitled to continue in possession under a statutory periodic tenancy (section 5 Housing Act 1988).

Where rent was payable under the original fixed term tenancy by reference to a period the periodic tenancy will be of that period. If no rent was payable the period of the tenancy will be monthly beginning on the day following the termination of the original licence or tenancy.

It is an implied term of any periodic assured tenancy that the tenant will not assign, sublet or part with possession of the whole or part of the dwellinghouse without the consent of the landlord.

There is also an implied term that the tenant shall give the landlord access and all reasonable facilities to execute repairs.

If the original tenancy was a periodic tenancy, that tenancy cannot be terminated unilaterally by the landlord. No statutory periodic tenancy arises but the original periodic tenancy continues.

4.3.7 Rent control
Under the Housing Act 1988 there are very limited provisions for the control of rent. The rent can be increased where there is a periodic tenancy without a rent review clause or where a statutory periodic tenancy has arisen.

The procedure is that the landlord must serve a notice of increase to take effect at the beginning of a new period of the tenancy. The notice has to be in a prescribed form and must be of at least the minimum period. This is six months for a yearly tenancy, one month for a tenancy of less than one month and the full period in other cases. The proposed date must be not less than 12 months from the date on which the last increase took place (section 13 Housing Act 1988).

The tenant has the right to refer a proposed rent increase to the Rent Assessment Committee for adjudication at any time before it takes effect. If he does not do so the rent will be increased from the date stated in the landlord's notice.

4.3.8 Grounds for possession
The landlord can obtain possession only if he can establish one of the statutory grounds for possession of an assured tenancy.

The mandatory grounds are covered by the following headings:

(1) Landlord requires property as his only or principal home
(2) Possession required by mortgagee
(3) Holiday lettings

(4) Student lettings
(5) Lettings to ministers of religion
(6) Demolition or reconstruction of premises
(7) Devolution on death of tenant
(8) Arrears of rent

The discretionary grounds are:

(9) Suitable alternative accommodation
(10) Less than three months arrears
(11) Persistent delay in paying rent
(12) Breach of tenancy agreement
(13) Deterioration of dwellinghouse
(14) Nuisance or immoral user
(15) Deterioration of furniture
(16) Tenant or anyone acting on his behalf obtained tenancy by false statement

It will be noted that ground 16 is omitted from the list as it is not available. This is the ground which enables a landlord to obtain possession on cessation of employment. If this ground were available it would undermine the principle of protecting agricultural workers upon the termination of their employment. It is only in those very limited circumstances when the tenancy is vested in a person who does not come within the definition of agricultural worker that ground 16 will be available.

4.3.9 Rehousing provisions

The rehousing provisions of the Rent (Agriculture) Act 1976 are applied to assured agricultural occupancies

4.4 The Rent Act 1977

A farmworker with a tenancy granted before 15 January 1989 who pays rent in excess of two-thirds of the rateable value will be protected by the Rent Act 1977. Generally speaking such a tenant will be in no different position from any other Rent Act protected tenant.

However, a landlord who has need of the property to rehouse an agricultural worker is able to apply to an ADHAC under section 27(2) of the Rent (Agriculture) Act 1976 (see also section 4.2.10.2).

Moreover, the landlord may be able to obtain possession under two grounds in Schedule 15 of the Rent Act 1977 which are not available to other landlords. These grounds will not however apply where the letting is to an agricultural worker who would have been protected under the Rent (Agricultural) Act 1976 were it not for the low rent.

Ground 16 gives a mandatory ground for possession where a house was at one time occupied by a worker in agriculture and the present tenant has never been employed by the landlord. Notice in writing must have been given before the start of the tenancy that possession might be recovered under this case for the housing of an agricultural worker.

Ground 18 applies where the former occupier was responsible for farming the holding of which the dwelling forms a part. If notice was given before the start of a letting to a non-agricultural tenant that possession might be sought on this ground, possession can be obtained for an agricultural employee.

4.5 The Protection From Eviction Act 1977

A farm worker who is unable to satisfy the agricultural worker condition contained in either the Housing Act 1988 or the Rent (Agriculture) Act 1976 and who is unable to claim the protection of the Rent Act 1977 or the protection given to tenants of assured tenancies under the Housing Act 1988 will enjoy the protection provided by the Protection From Eviction Act 1977. Section 4 of the Act provides that a farm worker who has occupied premises under the terms of his employment in agriculture has six months' security of tenure as from the date when his contractual right to occupy the dwelling under the service agreement ceases. That will usually be the date when his contract of employment terminates. However, the owner can apply to the county court for possession within that six month period and the court can make a possession order. But if he does so, the court must suspend the order for the remainder of the six month period unless it is satisfied that

(1) other suitable accommodation is, or will within that period, be made available to the occupier; or
(2) the efficient management of any agricultural land or the efficient carrying on of any agricultural operations would be seriously prejudiced unless the premises are available for occupation by a person to be employed by the landlord; or
(3) greater hardship would be caused by the suspension of the order than by its execution; or
(4) the occupier of the dwelling or anyone residing with him has been causing damage to the premises or has been guilty of misconduct which is a nuisance or annoyance to persons occupying other premises;

and being satisfied in one or other of these points, the court considers it would be reasonable not to suspend the execution of the order for the remainder of that period.

4.6 Shorthold tenancies

4.6.1 Purpose
Because of the difficulty a landlord had in gaining possession from a tenant who was either a protected agricultural occupant or an assured

agricultural occupant, many landlords let property to farm workers on shorthold tenancies. After some initial fears that this might be contrary to official policy it was generally accepted that these lettings were valid.

4.6.2 Protected shorthold tenancy

These were introduced by the Housing Act 1980. The tenancy had to be between 1 and 5 years and provided the appropriate notice was given at the start of the term possession could be obtained by the landlord under Case 19 of the Rent Act 1977.

No protected shorthold can be granted after 15 January 1989.

4.6.3 Assured shorthold tenancy

These were introduced by the Housing Act 1988. Before 24 February 1997, the commencement date of the Housing Act 1996, the tenancy had to be granted for a definite term of at least 6 months but there was no longer any need for a maximum term (section 20 Housing Act 1988). It was also essential to serve a notice in prescribed form before the start of the tenancy informing the tenant that the tenancy was to be an assured shorthold tenancy. These requirements have been abolished by the Housing Act 1996 for most assured shorthold but notices are still necessary if an assured shorthold is to be granted to an agricultural worker (section 96 and Schedule 7 Housing Act 1996). Although the tenancy no longer has to be for a minimum of six months, any order for possession may not take effect earlier than six months after the beginning of the tenancy.

It should be remembered that an assured shorthold is a variant of the assured tenancy and therefore the rent must be not less than £251 per annum.

A tenant of an assured shorthold may demand from his landlord a written statement of the terms of the tenancy, unless he already has such a statement (section 97 Housing Act 1996).

4.6.4 Rent control

A shorthold tenant can refer the rent to a Rent Assessment Committee if he considers that the rent is significantly higher than the rent payable under similar tenancies of similar dwellinghouses in the locality (section 22 Housing Act 1988). The Committee has power to fix the rent at a market rent taking into account the lack of security of tenure. The tenant's right to go to the assessment committee arises only during the initial fixed term. Moreover, the committee can only make a determination if there are a significant number of assured tenancies of similar dwellings in the locality.

Where the original fixed term tenancy continues as a periodic tenancy the rent may be increased by the landlord serving a prescribed notice on the tenant. If the tenant does not agree the proposed increase he can refer the rent to the Rent Assessment Committee (section 13 Housing Act 1988).

4.6.5 Security of tenure

The landlord may obtain possession on giving two months notice in writing to expire at the end of the fixed term or at the end of a completed statutory periodic tenancy (section 21 Housing Act 1988 as amended by Housing Act 1996). Where the periodic tenancy is quarterly, six monthly or yearly the notice must also accord with the common law rules. Therefore the notice period would be a quarter or six months.

The court has no discretion to refuse possession though in cases of exceptional hardship it may postpone possession for up to six weeks (section 89 Housing Act 1980).

Chapter 5
Business Tenancies on Farms and Estates

5.1 Introduction

Landowners may be able to let redundant farm buildings for commercial purposes such as offices, workshops, or small retail units. Before doing so they must obtain the necessary planning permission and be sure that there are no restrictive covenants in their deeds precluding commercial use. A building which is not used for agriculture will be liable to non-domestic rates. Any new letting will be governed by the Landlord and Tenant Act 1954 Part II. Where there are existing agricultural tenancies the position depends on whether the tenancy is governed by the Agricultural Holdings Act 1986 or the Agricultural Tenancies Act 1995.

5.2 Agricultural Holdings Act tenancies

The Agricultural Holdings Act 1986 applies to tenancies granted before 1 September 1995 and to succession tenancies granted after that date. The definition of an agricultural tenancy in the Agricultural Holdings Act, and the case law on this and previous Acts means that a tenancy cannot be partly governed by the agricultural holdings legislation and partly by the business code. Where the predominant use is agricultural then the tenancy will continue to be covered by the Agricultural Holdings Act. Where, however, there is a major shift away from agriculture then the letting will be governed by the Landlord and Tenant Act 1954 Part II.

A tenant who diversifies may find that he is in breach of some of the covenants relating to agriculture in his existing tenancy. An obvious example is a covenant to farm in accordance with the rules of good husbandry. It is therefore in his interest that his proposals for diversification should be discussed with the landlord.

Often the most sensible arrangement will be for the tenant to surrender that part of his premises he intends to use for the new enterprise. The landlord can then grant him a new lease, specifically geared to the business use of those premises. The tenant will remain an agricultural tenant on the rest of the property.

5.3 Farm business tenancies

A letting of agricultural land after 1 September 1995 will generally be a farm business tenancy as described in Chapter 2.

Although at the beginning of the tenancy the character of the tenancy must be primarily or wholly agricultural it is possible for the tenancy to remain a farm business tenancy, as distinct from a commercial business tenancy, even if there is a major diversification into non-agricultural use. Certain conditions must be fulfilled including the service of a notice before the beginning of the tenancy. The notice conditions are explained at section 2.2.3.

5.4 New tenants

If a landowner wants to let to a new tenant he must consider what terms are appropriate for the letting, including length of term and rent. Where it is proposed that the letting is to be for a long term or it covers more than the odd barn or shed it would be sensible to go to an agent or a solicitor to discuss the whole proposal with him. It may seem an unnecessary expense but it will be well worth it in the end to have properly drafted documents which deal with the particular concerns of the parties.

Some landowners may want to rely on their agents to provide suitable documents. It should be remembered, however, that it is an offence for a person who is not a solicitor or barrister directly or indirectly to prepare a lease, other than a farm business tenancy, for over three years. Leases under three years do not have to be by deed to transfer the legal estate but those of three or more years do. No offence is committed if the document was drawn up without any fee, gain or reward, but it is considered that a resident land agent would not come within the exception.

5.5 The legal nature of a lease or tenancy

A person who is given, or in practice takes, exclusive possession of property will have a tenancy. The document may be called a licence, or indeed the parties may have intended it to be a licence, but as a matter of law it will be a tenancy. In order to create a licence the landowner must continue to use the property for his own purposes as well as giving, the licensee certain rights.

At common law a fixed-term tenancy will end automatically at the expiry of the fixed term. A periodic tenancy has to be determined by the appropriate notice to quit. A weekly tenancy needs a week's notice, a monthly tenancy a month, a quarterly tenancy a quarter and a yearly tenancy six months. The notice must expire at the end of the completed period.

The term of a tenancy cannot be backdated. It will run from the date when the lease is executed (signed by the parties with the intention of it taking effect). Certain contractual obligations, for example to pay rent, can become due from an earlier date.

Where a tenant is let into occupation pending the conclusion of a formal lease he will have a tenancy at will. For the avoidance of doubt as to the exact arrangement the agreement should be in writing. Provided that it is clearly understood that the tenant is let in under those conditions then the tenancy can be terminated by either party if the negotiations break down. The tenancy will not be protected by the Landlord and Tenant Act 1954 Part II (see section 5.6). Unfortunately the position is not the same where an agricultural tenant was let into possession pending a formal lease before 1 September 1995. Section 2 of the Agricultural Holdings Act 1986 turned the tenancy at will into a fully protected agricultural tenancy.

5.6 The Landlord and Tenant Act 1954 Part II

5.6.1 Application of the Act
The Act

> 'applies to any tenancy where the property comprised in the tenancy is or includes premises which are occupied by the tenant and are so occupied for the purposes of a business carried on by him or for those and other purposes'.
>
> Section 23(1) Landlord and Tenant Act 1954

Business is widely defined. It 'includes a trade, profession or employment and includes any activity carried on by a body of persons whether corporate or unincorporate'.

The tenant does not have to occupy the whole premises to qualify for protection. Nor does he have to use the premises exclusively for business purposes. Premises used for mixed purposes are within the Act. So a cottage could be included with a workshop and the whole treated as a business tenancy provided the occupation of the cottage is incidental to the business use and not a sham; in these circumstances the cottage would not be covered by the Housing Act, 1988.

'Premises' can include bare land such as gallops for training racehorses (*Bracey* v. *Read* [1963] Ch 88).

The Act does not apply to

(1) Licences
(2) Incorporeal hereditaments (e.g. easements, profits)
(3) Agricultural Holdings Act tenancies
(4) Farm business tenancies
(5) Mining leases
(6) Leases of licensed premises
(7) Service tenancies
(8) Short fixed term tenancies of less than six months provided the tenancy does not contain provisions for renewing the term or extending it beyond six months or the tenant and any predecessor in

his business have not been in occupation for more than 12 months in total.

5.6.2 Termination by the landlord

Under a commercial tenancy the tenant has security of tenure so that the tenancy continues after the original fixed term has expired unless the agreement is brought to an end in accordance with the provisions of the Act. The only exception to this is where the tenancy or a superior tenancy is forfeited under a provision in the tenancy agreement.

In order to bring a tenancy to an end, the landlord must give between six and twelve months notice expiring on or after the contractual term date. The notice must be given on a form prescribed under the Landlord and Tenant Act 1954, Part II (Notices) Regulations 1983 SI 1983/133 and a landlord who does not want the tenancy to continue must state the grounds on which he opposes the grant of a new tenancy. As the time limits and procedures are strict a landlord should employ a solicitor who is well versed in the law of landlord and tenant to act for him.

A tenant has two months from receipt of the landlord's notice in which to serve a counternotice stating that he is unwilling to give up possession. If he does not serve the counternotice then he will have no right to apply for a new tenancy. His existing tenancy will end on the date specified in the landlord's notice and he must leave.

After a tenant has served a counternotice he must apply to the court for a grant of a new tenancy. The application must be made between two and four months after the service of the landlord's notice.

5.6.3 Termination by the tenant

The tenant can determine the tenancy by ordinary notice if it is a periodic tenancy. A fixed-term tenancy can be brought to an end by the tenant giving three months notice in writing to expire at the end of the term, or on any quarter day thereafter. If the tenancy includes break clauses, the tenant may serve notice and end the agreement at such times.

Alternatively, the tenant who wants a new tenancy to replace the existing one can serve on the landlord a request in statutory form for a new tenancy. This must specify a date for the start of the tenancy not less than 6 nor more than 12 months ahead. It must not be on a date before the existing tenancy would expire or could be determined. The tenant then has to apply to the court for the new tenancy not less than two nor more than four months after he has served the request on the landlord. The right to initiate a request for a new tenancy only exists where the tenant has a fixed-term tenancy exceeding a year or where the tenancy is for a term certain and then from year to year.

5.6.4 Grounds for opposing a new tenancy

Provided the landlord has specified one or more of the grounds in the statutory notice to terminate the tenancy he can seek to prevent the grant of a new tenancy to a tenant. The grounds available to the landlord are as follows.

(1) Breach of the tenant's repairing obligations.
(2) Persistent delay in paying rent.
(3) Substantial breaches of the tenant's other obligations, or any other reason connected with the tenant's use or management of the premises.
(4) The provision of suitable alternative accommodation.
(5) Where the current tenancy is a subletting of part only of the property comprised in a head lease and the reversioner of that head lease can demonstrate that the property could be let more economically as a whole.
(6) That the landlord intends to demolish, reconstruct or carry out substantial works of construction.
(7) That the landlord intends to occupy the premises for the purposes of a business carried on by himself (or of a company which he controls) or as his residence. This ground is not available to a landlord who has purchased his interest in the property within the preceding five years.

5.6.5 *The new tenancy*

If the landlord is not successful in his opposition to the grant of a new tenancy the court can grant a tenancy for a period not exceeding 14 years. The terms of the tenancy will be those agreed by the parties or in default of agreement those imposed by the court. The court will have regard to the terms of the former tenancy and all the relevant circumstances.

Once the order is made the tenant has 14 days in which to refuse to take the tenancy. The court must then revoke the order. It has, however, a discretion to order that the former tenancy shall continue for a period which will allow the landlord an opportunity to re-let.

5.6.6 *The rent*

The landlord and tenant can fix a market rent for the premises. A short lease may have no rent review provision but a well drafted longer lease should contain a rent review clause which will operate at stated times throughout the lease. Where the original fixed term has expired or where there is a periodic term the landlord may wish to increase the rent. In order to do so he must serve a notice to terminate the existing tenancy. This will set in motion the statutory renewal process which will result in the tenant being granted a new tenancy at the current market rent.

5.6.7 *Interim rent*

Once a landlord has served a notice to terminate the existing tenancy (or a tenant has served a request for a new tenancy) the landlord can apply to the court for the payment of an interim rent. Where the rent is expected to be higher than the existing rent it may be worth making an application but the interim rent will be lower than the full market rent which will be charged under the new tenancy. The new rent will run from the date of the application to fix it, or the date specified for the end of the former tenancy, whichever is later.

5.6.8 *Compensation for disturbance*

The tenant is entitled to compensation where the court refuses to grant a new tenancy on the grounds that the property could be let more profitably as a whole or that the landlord intends to develop the property or occupy it himself (see section 5.6.4, points (5), (6) and (7) in the list). These are all situations where there has been no fault on the part of the tenant. The tenancy is terminated for the benefit of the landlord.

The tenant is also entitled to compensation where the above grounds are specified in the landlord's notice and as a consequence the tenant does not make, or withdraws, his application for a new tenancy.

The landlord must pay the tenant a multiple of the rateable value as set out currently in the Landlord and Tenant Act 1954 (Appropriate Multiplier) Order 1990 SI 1990/363. For most cases, the multiplier is 1 making the compensation equal to the rateable value. Where the business has been carried out in the premises for the whole of the 14 years preceding the termination of the current tenancy the compensation is twice the rateable value.

5.6.9 *Compensation for improvements*

Unless there is a covenant in the lease preventing the tenant from making improvements he may be able to claim compensation on the termination of his lease. The law is contained in the Landlord and Tenant Act 1927. Agricultural holdings are expressly excluded from the provisions of the 1927 Act as are premises let to a tenant by virtue of his office, appointment or employment where the tenancy is in writing and expresses the purpose for which it was created.

The Landlord and Tenant Act 1927 does not lay down a list of tenant's improvements for which the landlord must pay compensation at the end of the tenancy. There is nothing in it to compare with the list of agricultural improvements for which compensation is payable under the Agricultural Holdings Act 1986. However, the following points should be noted:

(1) An improvement must add to the letting value of the holding.
(2) An improvement does not include any trade or other fixture which the tenant is by law entitled to remove. Generally, what is fixed to the property becomes part of it and cannot be removed. But 'trade fixtures' (e.g. petrol pumps, partitions) can be removed provided it is not forbidden by the terms of the lease.

(3) An improvement must be 'reasonable and suitable to the character of the premises'.
(4) An improvement should not diminish the value of any other property belonging to the same landlord.
(5) A claim cannot be made for an improvement made by the tenant where he has entered into a contract to make the improvements and received valuable consideration.

If a tenant is to be successful in a claim for compensation he must serve notice on his landlord of the intended improvement together with specification and plan. The landlord has three months to object. If the landlord does not object the tenant can proceed with the improvement. If he does object the tenant can apply to the county court or the Chancery Division of the High Court. The landlord can elect to do the improvement himself. If he does so he will not have to pay compensation to the tenant and he will be entitled to a reasonable increase in rent by agreement or as determined by the court. Where the tenant carries out the improvement, it must be made within the agreed time. A tenant who has duly completed an improvement may require his landlord to give him a certificate to that effect. If the landlord does not give a certificate then the tenant can apply to the court.

Sometimes improvements have to be made to fulfil statutory requirements. The procedure described above will apply in such cases except that the landlord cannot object to the improvement. Compensation is payable at the end of the tenancy for such improvements.

The claim for compensation must be made within three months of the first action taken to determine the tenancy, which may be the service of a notice to quit or under the Landlord and Tenant Act 1954 or be effected by forfeiture or re-entry. Where the tenancy expires at the end of a fixed term, the claim must be made not more than six nor less than three months before the termination. The claim must be in writing, signed by the tenant, his solicitor or agent, and must give a description of the premises, the trade or business and a statement of the nature of the claim, the particulars of the improvement and date of completion, together with the cost and amount claimed.

Generally, a tenant must quit the premises to obtain compensation. Disputed cases have to be referred to the county court or the High Court for settlement, depending on the rateable value of the holding.

Only two rules are laid down for assessing compensation (see section 1 (1) of the Landlord and Tenant Act 1927). They are as follows:

'The sum must not exceed
(a) the net addition to the value of the premises directly resulting from the improvement; or
(b) the reasonable cost of carrying out the improvement at the termination of the tenancy less the cost of putting the works into a

reasonable state of repair except so far as it is covered by the tenant's repairing liabilities.'

Under (a) above, account must be taken of the proposed use of the premises after termination of the tenancy, and if demolition, alteration or change of use is proposed, that is to be taken into account also. In these circumstances, an improvement can be worth nothing.

5.6.10 *Contracting out*

Section 38(4) of the 1954 Landlord and Tenant Act makes it possible to contract out of those parts of the Act, that give the tenant security beyond the fixed term. These rights are set out in sections 24–28 of the Act. A joint application by the intended landlord and intended tenant must be made to the court *before* the tenancy is granted. Sections 24–28 will not be excluded if the lease or underlease is executed before the court's authorisation is given unless the grant is conditional on that authorisation being given. In *Essexcrest* v. *Even* (1988) 55 P&CR 279 a unilateral endorsement of a lease after execution with a memorandum of the court's order was held to be ineffective.

No personal attendance is required and an order is usually made within a few days. Where the application is signed by individuals, and not their solicitors, the court will want to be sure that the tenant is properly aware of the implications of the order. The landlord should advise the tenant to take his own independent advice. Where the tenant does not do so he should be asked to sign a letter, stating that he has received that advice and is fully aware that his tenancy will terminate at the end of the fixed term and that he has given up his rights under the 1954 Act.

5.7 Stamp duty

Stamp duty may be payable on the lease depending on the amount of the premium, if any, the length of the term and the rent payable. The lease, which is the document signed by the landlord and held by the tenant, and its duplicate (or counterpart) should be lodged for stamping with duty at the appropriate rate within 30 days of the grant. It is for the tenant to arrange for the stamping of the lease and he should be advised of his obligation. The duplicate or counterpart should be stamped by the landlord. At the time of writing the duty is fixed at £5.00.

Leases which are liable for stamp duty but have not been stamped cannot be produced in court. It is possible for documents to be stamped late on the payment of a penalty.

If a lease is granted for seven years or more, it fails within section 28 of the Finance Act 1931 and must be produced to the Inland Revenue in accordance with the requirements of that section.

The rules on, and rates of, stamp duty change from time to time. The current position and more detailed advice should be obtained from the Office of the Controller of Stamps, Inland Revenue, South West Wing, Bush House, Strand, London WC2B 4QN.

5.8 Registration

If a lease is granted for over 21 years, the title to the lease must be registered at the Land Registry if the lessee is to secure legal title.

It must be stressed that it is important for both landlords and prospective tenants to be professionally advised where leases of this length are being prepared.

Chapter 6
Public Access

6.1 Introduction

The public have a right of passage along highways and will have a right to wander over mapped open country and registered common land granted by the Countryside and Rights of Way Act 2000 (see section 6.4.1). Common land was registered under the Commons Registration Act 1965 where it was subject to rights of common or was waste or former waste of a manor. The registration of a common is conclusive evidence that it is a common even if the registration was made in error. The public also have a right of access to town and village greens.

6.2 Public rights of way

6.2.1 Definition

To most people the word 'highway' conjures up a picture of a metalled road with motor cars speeding along it but a dusty footpath across a field is also a highway.

A highway is a public right of way over a defined linear route. The terms 'highway' and 'public right of way' are used interchangeably although sometimes the term highway is used to denote the physical route. Roads normally used by motor vehicles are seldom described as public rights of way. The legal consequence of the land being a highway is that the public have a right to pass and repass along the route. Any other activity, unless it is incidental to a right of passage will be a trespass.

6.2.2 Ownership of highway

Where a highway is maintainable at public expense the surface of the highway is vested in the highway authority. The landowner will retain ownership of the subsoil. Deciding whether a highway is publicly maintainable is complex. But as a guiding principle it will be so if it has been created formally or was in existence on 16 December 1949. If not repairable by the highway authority it may be repairable by a private landowner (for example under the provisions of an enclosure award) or by no one at all.

6.2.3 Recording of highways

The Wildlife and Countryside Act 1981 provides for the recording of certain highways on a definitive map held by county councils. It should be stressed that the procedures under this Act are to ensure that existing rights are properly recorded. New rights are not created.

There are four main categories of rights of way which are recorded on this map, namely a *byway open to all traffic* (BOAT), a *restricted byway*, a *bridleway* and a *footpath*.

A BOAT

'means a highway over which the public have a right of way for vehicular and all other kinds of traffic, but which is used by the public mainly for the purposes for which footpaths and bridleways are so used'.

section 66(1) Wildlife and Countryside Act 1981

A restricted byway is a new category of right of way introduced by the Countryside and Rights of Way Act 2000. When section 47 of the Act comes into force, all routes recorded on the definitive map as roads used as public paths (RUPP) will automatically be redesignated as restricted byways. On restricted byways the public will have a right of way

- on foot;
- on horseback or leading a horse; and
- in vehicles other than mechanically propelled vehicles.

These are the minimum rights. A restricted byway may also include the right to drive animals. The public may also be able to establish on the basis of historic evidence that there is a right for mechanically propelled vehicles.

On a bridleway the public have a right of way on foot and on horseback or leading a horse and on a footpath a right of way on foot. In both cases the rights are without prejudice to any higher rights which may be proved.

By virtue of section 30 of the Countryside Act 1968 the public may ride a pedal cycle on a bridleway. However, cyclists must give way to pedestrians and horse riders and are subject to any byelaws made by the local authority. There is no obligation to make up a bridleway so that the surface is suitable for cyclists.

Where additional paths or higher rights are claimed an application may be made to the surveying authority for a modification order to the definitive map. Applications may also be made to delete or downgrade paths where they are wrongly recorded. When making the order the council is only concerned with evidence of the existence or non-existence of the rights. Matters of amenity, security or suitability are irrelevant.

The making of a modification order is the first step in the process. If no objections are received to the order the council may confirm the order. If there are objections the order must be referred to the Secretary of State for the confirmation of the order or otherwise. Generally he will appoint an inspector to hold a public inquiry when representations may be made by all those objecting to the order.

The Countryside and Rights of Way Act 2000 has introduced a cut off date of 2026 for recording footpaths and bridleways that existed before

1949 but are not recorded on definitive maps. In addition, it will not be possible after that date to upgrade recorded footpaths, bridleways or restricted byways to show higher rights created before 1949. Conversely it will not be possible to downgrade footpaths wrongly recorded as bridleways.

6.2.4 Public path orders

Public path orders are made under the Highways Act 1980. The orders may create new paths, extinguish or divert existing ones. There is a legal adage 'once a highway always a highway'. This means that if there was a highway in the past it will continue to have legal existence, even if it has not been used for years and years and even if it is not visible on the ground, unless it has been formally extinguished by order

It is very difficult to obtain an extinguishment order. To do so it has to be shown that the path is not needed for public use. As soon as an application is made for an extinguishment order it is likely that people will claim that the path is used regularly or the only reason that it is not is because it has been blocked.

A farmer is better advised to apply for a diversion order. Before doing so he should discuss the proposed diversion with the parish council and any local interest groups so as to avoid opposition to the order. In the past where there has been opposition some local authorities have refused to make an order. The Countryside and Rights of Way Act 2000 has improved the position of owners, lessees and occupiers of any land used for agriculture, forestry or the breeding or keeping of horses. They are given a formal right of application for a diversion order with a right of appeal to the Secretary of State against an authority's refusal to make an order.

Before making an order the highway authority may ask the applicant to pay the administrative costs of the procedures. He may also be required to pay compensation to a third party over whose land the new diverted route will run and also the expenses incurred in making up the path.

Where an order is made to which there are objections the order is referred to the Secretary of State for confirmation or otherwise. It is usual for him to appoint an inspector to hold a public inquiry.

A diversion order can be an expensive and slow business taking up to two or more years. Therefore if there are other means of solving perceived problems they should be considered.

6.2.5 Dedication and acceptance

A highway may be created at common law by dedication by the landowner and acceptance by the public. Although dedication may be express it is usually inferred by long user by the public. The user is evidence of a presumed dedication and is also evidence of public acceptance. The user must be without force, secrecy or permission and must not be attributable to the tolerance of the landowner. The length of user which is required depends on the particular circumstances. In many cases no dedication can be inferred because there was no one with the capacity to dedicate a right of way.

It was to deal with these problems that a statutory presumption of deemed dedication was introduced in 1932. It is now in section 31 of the Highways Act 1980. Where there is 20 years use without force, secrecy or permission there will be a deemed dedication unless the landowner can show an intention not to dedicate. He may do this by closing the path for one day a year, by displaying notices making it clear that there is no public right of way, or by turning people back. The period of 20 years unencumbered user is the 20 years before any challenge to that user is made.

The most effective method is for the landowner to deposit a statement and map with the highway authority showing what ways he admits as public rights of way and lodging statutory declarations at intervals of not more than six years (the Countryside and Rights of Way Act 2000 will when the appropriate section is in force increase this to ten) stating that no other ways have been dedicated.

Where the land is let the landlord may exhibit a notice on the land rebutting any dedication. Those with interests under strict settlements may take proceedings for trespass or an injunction as if they were in possession.

6.3 Limitations on land use caused by public rights of way

6.3.1 *Obstructions*

A landowner must not commit a highway nuisance or obstruct a highway.

A public nuisance at common law in relation to a highway is 'any wrongful act or omission upon or near a highway, whereby the public are prevented from freely, safely and conveniently passing along the highway.'

An obstruction is 'anything which substantially prevents the public from having free access over the whole of the highway which is not purely temporary in nature.'

Under section 137 of the Highways Act 1980 a person who without lawful authority or excuse wilfully obstructs a highway is guilty of an offence and is liable to pay a fine. The Countryside and Rights of Way Act 2000 has introduced an additional remedy under section 39, whereby a person who has been given notice to remove an obstruction twice within three years is guilty of an offence.

A magistrates court on the conviction of a person for wilful obstruction of the highway may order that person to remove the obstruction, provided it is in that person's power to do so. This may be an addition to, or instead of, a fine. The order must stipulate what steps are required to remove the obstruction and the timescale within which it must be done. The time allowed by the order may be extended on an application made before the time runs out.

Failure to comply without reasonable excuse with an order to remove an obstruction is an offence punishable by a fine up to level 5 on the standard scale. If the offence is continued the convicted person is guilty of a further offence and is liable to a fine not exceeding one twentieth of that level for each day on which the offence is continued. No offence or further offence is committed during any time given by the court, to the person against whom the order has been made, to effect the removal.

Where a person has been convicted of failure to remove an obstruction and the highway authority removes the obstruction the authority can claim from that person its expenses in doing so.

6.3.2 Stiles, gates, fences, wires

Structures across a public right of way such as stiles, gates, fences and wires amount to an obstruction unless the path was originally dedicated subject to them. If a way has not been dedicated subject to gates or stiles, a statutory notice may be served on the occupier requiring him to remove the structure. If he fails to do so, the authority may remove it and recover its reasonable expenses. In addition, the highway authority may prosecute for obstruction.

Gates in carriageways and bridleways are required to be of a minimum width. For a bridleway the minimum width is five feet, for a carriageway, including a BOAT the minimum width is ten feet, measured between the posts. Where the gate is less than the minimum width, the highway authority may by notice to the owner of the gate require him to enlarge it to that width or remove it. Failure to do so within 21 days of service of the notice is an offence.

In limited circumstances the highway authority may give permission to an owner, lessee or occupier for the erection on footpaths or bridleways of a gate or stile to prevent ingress or egress of animals on land which is used, or is being brought into use, for agriculture or forestry or for the breeding and keeping of horses. The authority has no power to authorise gates or stiles for other purposes such as security. Conditions may be imposed on the size and type of barrier.

The maintenance of a stile, gate or other structure is generally the responsibility of the landowner subject to a contribution of not less than one quarter of any expenses from the highway authority. Liability for maintenance may be wholly vested in the highway authority or district council if the landowner can obtain agreement to this course.

If a landowner does not maintain a stile in a safe condition and to the standard of repair required to prevent unreasonable interference with the rights of persons using the footpath or bridleway, the highway authority or district council, after giving 14 days notice, may enter to do the work and recover the whole of the cost.

Fences across a path, unless authorised, will be an obstruction. Electric or barbed wire fences along the side of the path may amount to a public nuisance. If members of the public using the path are likely to wander into the fence, bearing in mind that the path may be used at night, the fence will amount to a nuisance.

As well as being a nuisance at common law, a barbed wire fence can be the subject of a statutory notice under section 164 of the Highways Act 1980. The occupier of the land may be required to remove the fence within a specified time, and if he fails to do so, a complaint made be made to a magistrates' court for the making of an abatement order. If the order is not complied with the authority may do the work in default and recover the expenses.

It is an offence under the Highways Act 1980 to place a rope or wire across a highway in such a way as to cause danger to a user of the highway. This is sometimes done on public paths to control stock temporarily, but it can be dangerous to horse riders, pedestrians and farm traffic which may have a private right of way along the track.

6.3.3 *Overhanging vegetation*
Vegetation which overhangs the sides of a path may obstruct the passage of vehicles or pedestrians or endanger or obstruct the passage of horse riders. Where it does so the highway authority may serve a notice on the owner of the tree, shrub or hedge or on the occupier of the land on which it is growing, to cut back or lop the vegetation so as to remove the cause of danger, obstruction or interference. Any person aggrieved by the notice may appeal to the magistrates' court. If the works are not carried out in accordance with the notice the authority can do the work itself and recover the expenses. This power is available whether the vegetation has grown naturally or been deliberately planted.

6.3.4 *Crop spraying*
The spraying of crops can interfere with the use of a right of way. It can also endanger people and animals and give rise to an offence under the Health and Safety at Work Act 1974.

If the product label states that people and animals should stay out of a treated crop, warning signs should be placed where the paths enter the sprayed area. The signs should state something to the effect of 'Sprayed: Please Keep to the Path'. They should be left in place until it is safe.

Spraying should be temporarily stopped if anyone is using a path which crosses or adjoins the field which is being sprayed.

Any herbicides used to deal with vegetation should be applied in accordance with the various health and safety regulations and any current DEFRA codes of practice.

6.3.5 *Ploughing and other disturbances*
The purpose of the Rights of Way Act 1990 was to clarify the rights and responsibilities of farmers and local authorities where paths were disturbed by agricultural activity. Its aim was to strike a balance between the need to farm without undue hindrance and the right of path users to enjoy

the countryside while at the same time giving local authorities a clearer framework within which to operate. The previous provisions giving a statutory right to plough together with the ploughing code had not proved satisfactory in achieving these objectives.

Ploughing a public right of way is an interference amounting to a public nuisance at common law unless a right of way was dedicated subject to the right to plough. If this right is recorded in a statement accompanying the definitive map, it is conclusive evidence of its existence. Otherwise the onus is on the farmer to show that he has a common law right to plough. In most circumstances the farmer would be better advised to rely on his statutory right to plough.

The statutory right to plough or otherwise to disturb the surface of a footpath or bridleway is contained in the Highways Act 1980 as amended by the Rights of Way Act 1990. The statutory right applies only to crossfield footpaths and bridleways over agricultural land. There is no right to plough field edge paths. These are defined as footpaths or bridleways that follow the sides or headlands of fields or enclosures.

Byways open to all traffic may not be ploughed. Whether or not a RUPP (see section 6.2.3) may be ploughed depends on whether it has vehicular rights. Until a RUPP is re-classified under the Wildlife and Countryside Act 1981 it may not be clear what rights do exist on one. The presumption is that it is a bridleway. If it is alleged that there is no right to plough crossfield RUPP it must be proved that vehicular rights exist.

When section 47 of the Countryside and Rights of Way Act 2000 comes into force RUPPs will automatically become restricted byways. As a restricted byway has vehicular rights, albeit for non-mechanically propelled vehicles, there will be no right to plough under section 134 of the Highways Act 1980.

The right to plough or otherwise disturb the surface in accordance with the rules of good husbandry of crossfield bridleways and footpaths is permitted only where it is not reasonably convenient to avoid ploughing or disturbing the surface when sowing or cultivating a crop.

After ploughing or other disturbance the surface must be made good to at least the minimum width, so that the path is reasonably convenient to use and the line is apparent on the ground. This must be done within 14 days of the first disturbance for the sowing of the crop or of any subsequent disturbance in connection with that crop within 24 hours, unless a longer period has first been agreed in writing by the Highway Authority. Where an extension is required, an application must be made to the highway authority before the disturbance occurs or is in the relevant time. The extension is subject to a maximum of 28 days.

Crops, other than grass, will be an obstruction if they grow on or overhang the minimum width of any footpath bridleway or unmade carriageway so as to inconvenience the public or prevent the line being apparent on the ground. Unmade carriageways include BOATs and RUPPs with vehicular rights and restricted byways. A crop will be treated as grass only if 'it is of a variety or mixture commonly used for pasture, silage or haymaking, whether or not it is intended for such a use in that case and is not a cereal crop'.

There is no automatic right to carry out excavations or engineering operations on a public right of way. Where, however, such works are reasonably necessary for the purposes of agriculture the occupier may apply to the highway authority for an order allowing him to disturb the surface of a footpath or bridleway for the period specified in the order. The period is limited to a maximum of three months.

The highway authority must make the order if it is satisfied that it is practicable temporarily to divert the path in a manner reasonably convenient to users or that the existing path will be or can be made sufficiently convenient while the works are undertaken. When a diversion is made onto land not occupied by the applicant the written consent of the occupier of that land, and of any other person whose consent is needed to obtain access to it, must be obtained.

The order may be made subject to conditions for the benefit of users of the path and where there is a diversion, conditions relating to the provision of signposts and other notices, stiles, bridges and gates. The authority may recover from the applicant expenses in connection with the order.

The minimum widths for keeping paths clear are as follows.

(1) The width recorded in the statement attached to the definitive map, if there is such a width recorded (and there seldom is); or
(2) (a) a footpath: 1 m for a crossfield path, 1.5 m for a field edge path;
    (b) a bridleway: 2 m for a crossfield path, 3 m for a field edge path;
    (c) for any other right of way 3 m whether a crossfield or field edge.

If the occupier does not restore a path to the minimum width within the statutory time limits, the highway authority has powers of entry to do the necessary work and recover the cost. The authority must give not less than 24 hours notice before entering to do the work. No notice is needed if the entry is just for the purpose of obtaining information. The default powers may be exercised up to a maximum width. This is to provide some tolerance for different types of machinery and equipment and is intended to limit disputes about the exact width that has been restored or cleared. The maximum widths are as follows:

(1) The width (if any) recorded in the statement attached to the definitive map, if there is no such width recorded; or
(2) (a) a footpath: 1.8 m for a crossfield path and field edge path;
    (b) a bridleway: 3 m for a crossfield path and field edge path;
    (c) other unsurfaced highways: 5 m.

Only the highway authority or the council of the district, parish or community in which the offence is committed may bring a prosecution against the occupier for not complying with the ploughing or disturbance provisions of the Rights of Way Act 1990 as incorporated into the

Highways Act 1980. This limitation will be removed when section 70 of the Countryside and Rights of Way Act 2000 comes into force. There is no such limitation where an occupier fails to keep a public right of way clear of crops. Any person may bring a prosecution.

6.3.6 Dangers

Neither the highway authority nor the landowner is liable under the provisions of the Occupiers' Liability Acts 1957 and 1984 to users of the highway. This is because the public have a legal right to use the public right of way and neither the authority nor the landowner can exclude them.

Where, however, a landowner or occupier offers the public an alternative route, which is not an official diversion, the public are given a licence to use that route. The public are not using the route as of right. As they enter that part of the land by express or implied permission they become invitees. Therefore there is potential liability under the Occupiers' Liability Act 1957.

The position may be the same where, because of an obstruction caused by the landowner, the user deviates onto land in the ownership of the same landowner, on the basis that the landowner has given an implied permission for the deviation.

An occupier may be liable in nuisance if he does something which interferes with a member of the public's enjoyment of a right of way. An example would be depositing something on the path, or erecting a wire fence too close to a footpath.

6.3.7 Trees

A highway authority has the power to serve a notice on the owner or occupier of land in which a hedge, tree or shrub is growing requiring him to cut or fell it if it is dead, diseased, damaged or insecurely rooted and is likely to cause danger by falling onto the highway. There is a right of appeal against the notice to the magistrates' court. If there is no appeal and the work is not done the highway authority can do the work and recover the costs from the owner. The owner of a tree falling onto a public right of way which causes injury to a user is not liable for the damage unless negligence can be proved against him. Liability depends on whether the owner knew, or should have known, that the tree was likely to fall and cause injury. Owners of trees adjoining the highway should therefore have periodic inspections and fell trees where necessary. It is wise to insure against the risk of damage by trees falling onto the highway.

Highway authorities have powers to remove trees which are blown down across highways. The authority has powers to recover its reasonable costs for removal of fallen trees except where the owner can show that he took reasonable care to secure that the tree did not cause or contribute to the obstruction. So the owner of trees who has them regularly inspected and finds them healthy and safe would not be liable to pay the highway authority's expenses of removal should a tree fall over in a gale.

6.3.8 Shooting and firearms

It is an offence under section 161 the Highways Act 1980 to discharge any firearm or firework within 50 feet of the centre of a carriageway without lawful authority or excuse if as a result a user of the highway is injured, interrupted or endangered. This offence relates only to BOATs, restricted byways and RUPPs which have vehicular rights but not to footpaths or bridleways.

Although there is no specific offence of shooting across a footpath or bridleway it may amount to a public nuisance or to wilful obstruction of the highway. There may also be liability in negligence where it is known that people are on the path.

Section 19 of the Firearms Act 1968 makes it an offence for a person without lawful authority or reasonable excuse to have in a public place a loaded shotgun or loaded air weapon or any other firearm whether loaded or not together with ammunition. A public place is defined to include any highway and any other premises to which at the material time the public have or are permitted to have access.

The Town Police Clause Act 1847 prohibits the discharge of firearms in any street, so as to obstruct annoy or endanger residents or passers by. Although originally this provision applied only to those towns which adopted it, the provision was subsequently extended to all urban areas. Due to local government reorganisation the Act now applies to all England and Wales. Street is defined as 'any road, square, court, alley and thoroughfare or public passage'.

6.3.9 Fires

A person commits an offence if without lawful authority or excuse he lights a fire on a carriageway or on land adjoining the carriageway and as a result a user is injured, interrupted or endangered by the fire or the smoke.

6.3.10 Unfenced dangers

Where there is an unfenced or inadequately fenced source of danger near a public right of way the local authority may by notice on the owner or occupier of the land require him to do such works as will obviate the danger. The person who is aggrieved by such a notice may appeal to the magistrates' court. If he does not appeal and fails to comply with the notices within the specified time the authority may execute the works and recover the expenses.

Under the Mines and Quarries Act 1954 abandoned and disused mines must be fenced so as to prevent any person from accidentally falling down a shaft or entering an outlet. Failure to do so where by reason of its accessibility from the highway or a place of public resort such a mine constitutes a danger amounts to a statutory nuisance.

The same is true for quarries, whether being worked or not, where they are not provided with an efficient and properly maintained barrier so designed and constructed as to prevent any person from falling into them.

6.3.11 Animals

The keeping of animals on or near a public right of way may amount to an obstruction or highway nuisance if it prevents the free, safe and uninterrupted passage of the public.

There are special provisions in section 39 of the Wildlife and Countryside Act 1981 relating to bulls. The general rule is that it is an offence for the occupier of a field crossed by a right of way to keep a bull in a field. No offence will be committed if either the bull is less than 10 months old or is not a recognised dairy breed and is at large with cows or heifers. Dairy breeds are Ayrshire, British Friesian, British Holstein, Dairy Shorthorn, Guernsey, Jersey and Kerry.

This provision is not wholly satisfactory as many people are unable to make the distinction between dairy and beef cattle. However, it was a compromise between the farming community and the users of rights of way.

Even if the bull comes within the exception there may be liability under the Health and Safety at Work Act 1974. Section 3 places an obligation on employers and self-employed persons not to put at risk the health and safety of persons not in their employment.

A dog which prevents the use of a path by behaving in a threatening manner may constitute a public nuisance at common law. However, it has been held that there was no obstruction where a user was frightened by Rottweiler dogs in an adjoining garden putting their paws and muzzles on a fence adjoining a public footpath.

It is an offence under the Town Police Clauses Act 1847 in any street (defined to include any road, square, court, alley, thoroughfare or public passage) for a person to allow an unmuzzled ferocious dog to be at large or who sets on or urges any other dog or other animal to attack, worry or put in fear any person or animal.

The Dogs Act 1871 as extended by the Dangerous Dogs Act 1989 empowers a magistrates' court to order a dangerous dog to be kept under proper control or destroyed.

Under the Guard Dogs Act 1975 it is an offence to use a guard dog at any premises, other than a dwelling or agricultural land, unless tied up or in the control of a handler and a notice warning of the dog's presence is exhibited at the entrance of the premises.

The Dangerous Dogs Act 1991 makes it an offence if a dog is dangerously out of control in a public place and an aggravated offence if the dog while so out of control injures any person. 'Public place' means any street, road or other place which the public have or are permitted to have access. A dog is dangerously out of control on any occasion on which there are grounds for reasonable apprehension that it will injure any person, whether or not it actually does so.

The Animals Act 1971 imposes liability on the keeper of an animal belonging to a dangerous species or of an animal who although not of a dangerous species is known to have vicious tendencies. It is unlikely that animals of a dangerous species will be set free on public rights of way but animals such as horses and bulls, even though not prohibited, may be known to cause injury to passers-by. There will be no liability, however, if the damage was wholly the fault of the injured person.

In the unreported county court case of *Birch* v. *Mills* a solicitor was exercising his dogs on short leads on a public right of way. He was injured when a herd of Charolais cows charged at the dogs. A similar incident had occurred a few weeks earlier and had been reported to the farm manager. The defendant was therefore held liable.

Straying animals may amount to an obstruction of the highway. It is the duty of the owner of livestock to keep his animals fenced in. If the highway runs across land which is traditionally unfenced there is a special statutory provision. Section 8 of the Animals Act 1971 provides

> 'Where damage is caused by animals straying from unfenced land to a highway a person who placed them on the land shall not be regarded as having committed a breach of duty to take care by reason only of placing them there if –
> (a) the land is common land, or is land situated in an area where fencing is not customary, or is a town or village green, and
> (b) he had a right to place the animals on that land.'

Under section 155 of the Highways Act 1980 the keeper of any horses, cattle, sheep, goats or swine found straying on or lying on or at the side of a highway will be guilty of an offence. The offence will not be committed where the highway crosses common, waste or unenclosed land.

6.4 Access to open land and registered common land

6.4.1. The Countryside and Rights of Way Act 2000
This Act received Royal Assent on 30 November 2000. The access provisions are unlikely to be in force fully before 2004/5. The main access provisions under the Act are set out in the following sections.

6.4.2 Open land
The Act introduces a new right of access on foot for open air recreation to mountain, moor, heath, down (but not land which appears to the Countryside Agency or the Countryside Council for Wales to consist of improved or semi-improved grassland) and registered common land.

There is provision for the definition to be extended to include coastal land and for landowners to dedicate any land for public access.

Open country will be shown on maps to be prepared by the countryside bodies. However, land over 600 metres above sea level and registered common land will immediately qualify as open land, i.e. before it is shown on the maps. There is a procedure for publishing draft and provisional maps with a right of appeal by those with an interest in the land which they claim should not be included as access land. Access maps are to be reviewed at least every 10 years.

Land over which there is an existing right of public access (e.g. commons within section 193 of the Law of Property Act 1925 or land subject to access agreements or orders under the National Parks and Access to the Countryside Act 1949) will not count as access land for the purpose of the legislation. However, the existing right of access will continue.

6.4.3 Excepted land

Certain land, although it comes within the statutory definition of open land, will not be available for public access. The excepted land includes cultivated land, land covered by buildings or by pens for the temporary reception or detention of livestock, land used as a park or garden, or for the getting of minerals by surface working, or for the purpose of a railway or tramway, or a golf course, racecourse or aerodrome. Excepted land also includes land within 20 m of a dwelling or of a building used for housing livestock. In order to qualify as excepted land any necessary planning consents must have been obtained. Land over which there are in force byelaws made by the Secretary of State for Defence will also count as excepted land.

6.4.4 Exclusions and restrictions on access

Land owners will be able to exclude or restrict access for any reason for up to 28 days a year, but bank holidays may not be excluded and no more than four of the excluded days in any calendar year may be either a Saturday or Sunday. No Saturday in the period beginning with 1 June and ending with 11 August in any year may be excluded. No Sunday in the period beginning with 1 June and ending with 30 September in any year may be excluded. The purpose of these rules is to ensure that access is not precluded during the periods when the public are most likely to want to exercise the statutory right.

Further exclusions or restrictions may be approved by the Countryside Agency for land management reasons and on grounds of nature and heritage conservation, fire prevention, defence or national security, and to avoid danger to the public.

6.4.5 Exclusion of dogs

The owner of a grouse moor may exclude for a specified period, not exceeding five years, people with dogs if it appears necessary for the management of the land. The owner of land used for lambing may exclude people with dogs from any field or enclosure of not more than 15 hectares for one period of not more than six months in any calendar year. The steps which an owner must take before the exclusions apply will be specified in regulations.

The exclusions do not bar blind persons with trained guide dogs, or deaf persons with trained hearing dogs.

Dogs must be kept on a short lead, two metres or less, between 1 March and 31 July in every year and whenever they are in the vicinity of livestock.

6.4.6 Restrictions on the right of access

Where the right exists it is subject to various restrictions to control activities which are not compatible with the quiet exercise of the right. For instance vehicles, including bicycles are not allowed, nor water craft. Horse riding is not permitted. Other prohibited activities include unauthorised feeding of livestock, interfering with any safety fence or barrier, metal detecting and committing any criminal offence.

Those who do not observe the restrictions or who break or damage any wall, fence, hedge or gate will become trespassers and will lose their right of access to that land or land in the same ownership for 72 hours.

6.4.7 Rights and liabilities of owners and occupiers

The statutory right of access does not increase the liability of a person interested in the land in respect of its state or things done on the land. The exercise of the statutory right overrules any restrictive covenant on use.

A person who enters access land under the Act does not count as a visitor for the purposes of the Occupiers' Liability Act 1957. The occupier owes a duty of care, similar to that owed to trespassers under the Occupiers' Liability Act 1984. However, specifically excluded is any duty in respect of a risk resulting from a natural feature of the landscape or of any river, stream, ditch or pond, whether or not it is a natural feature. Plants, shrubs or trees, of whatever origin, are regarded as a natural feature of the landscape. Also excluded are risks of a person suffering injury when passing over, under, or through any wall, fence or gate, except by proper use of the stile or gate. However, the occupier will owe a duty where the danger is due to anything done by the occupier with the intention of creating the risk or being reckless as to whether that risk was created.

The duty arises if an occupier

(1) is aware of a danger or has reasonable grounds to believe that it exists;
(2) knows that a person is near a danger or may come near it; and
(3) the risk is one against which, in all the circumstances of the case, he might reasonably be expected to offer some protection.

The duty is to take such care as is reasonable in all the circumstances to see that the person does not suffer injury. The occupier may discharge his duty by giving warnings of the danger or discouraging people from taking the risk.

In determining whether and if so what duty is owed by an occupier of land where there is a right of access under the Countryside and Rights of Way Act 2000 regard is to be had to

'(a) the fact that the existence of that right ought not to place an undue financial or other burden on the occupier,
(b) the importance of maintaining the character of the countryside, including features of historic, traditional or archaeological interest, and
(c) any relevant guidance given under section 20.'
section 13(3) Countryside and Rights of Way Act 2000

Where a person is owed a duty under the Occupiers' Liability Act 1984 the occupier does not incur any liability for loss or damage to property. No duty is owed to any persons for risks willingly accepted.

A landowner will be liable if he displays a notice containing false or misleading information likely to deter the exercise of the statutory right.

6.4.8 Means of access

If access land has no legal right of access the Countryside Agency may make an application to the Secretary of State (or the Countryside Council for Wales may make an application to the National Assembly for Wales) for a creation order under section 26 of the Highways Act 1980.

There may also be a need for physical works to be carried out, such as the construction of gates or openings made in walls and hedges. Powers are given to access authorities (highway authorities or National Parks) to enter into agreements with landowners and occupiers to do the necessary works or where agreement cannot be reached to do the work themselves after giving the stipulated notice.

6.4.9 Management

There is a statutory duty for the countryside bodies to issue codes of conduct setting out rights and obligations and information in relation to access land.

The local highway authority (or where the land is in a National Park, the National Park authority) is under a duty to establish a local access forum which will consist of members appointed in accordance with regulations. The views of local access forums are to be taken into account in the preparation of maps of open country, making byelaws, appointing wardens, and before making restrictions and exclusions which may exceed six months.

Access authorities are given the power to make byelaws to preserve order, prevent damage, and to secure good behaviour so there is no undue interference with the rights of others. There is also a power to appoint wardens. Their job will to be to advise and assist the public and landowners, and to secure compliance with byelaws and the general restrictions.

Access authorities may put up notices indicating the boundaries of access and excepted land and informing the public of the general restrictions and any particular exclusions and restrictions.

6.5 Towns and village greens

Town and village greens are areas of land, registered under the Commons Registration Act 1965, which come within the following definition in section 22(1)

'(a) land which has been allotted by or under any Act for the exercise or recreation of the inhabitants of the locality
(b) land on which the inhabitants of any locality have a customary right to indulge in local sports and pastimes
(c) land on which the inhabitants of any locality have indulged in such sports and pastimes as of right for not less than twenty years.'

The phrase 'lawful sports and pastimes' includes such activities as the village cricket match, maypole dancing and practising archery.

The recent case of *R v. Oxfordshire CC ex parte Sunningwell Parish Council* [1999] 3 WLR 160 has made it much easier to register new town and village greens under (c) above. First, 'sports and pastimes' was held to include more informal activities such as blackberry picking, flying kites and tobogganing. Second 'as of right' merely means without force, secrecy or permission and does not require any subjective belief by the inhabitants of the existence of the right, nor is toleration by the landowner inconsistent with user as of right. Third, it was held to be sufficient if the land was used predominantly, rather than exclusively, by the inhabitants of the village.

The predominant test has been altered by the Countryside and Rights of Way Act 2000. Land will now qualify under (c) above if 'a significant number of the inhabitants of any locality, or of any neighbourhood within a locality' have used it for at least 20 years for lawful sports and pastimes. This removes the need to demonstrate that use is predominantly by people from the locality and means that use by people from outside that locality will no longer have to be taken into account by registration authorities. The amendment also deals with the problem of showing that users come from a particular village or parish. This is difficult in large built-up areas and so the concept of neighbourhood has been introduced.

Another difficulty has also been dealt with by the amendment. The application for the registration of a town or village green had to show 20 years user up to the date of the application. Often when the user is prevented by the landowner it takes a considerable time to collect the necessary evidence to make the application. By the time this is done there may be a gap in the user which would defeat the claim. Provision has therefore been made for the Secretary of State to make regulations setting a time limit within which an application to register must be made. The present government's (2001) intentions are that it should be two years. If no application has been made within two years of the land ceasing to be used for lawful sports and pastimes, the owner or developer will be able to take whatever steps are necessary to develop the land in the certainty that an application for the registration of a green will be rejected.

The new regime will cause difficulties for those wanting to develop land. There is a possibility that any vacant land might be registered even though the land does not have the appearance of a traditional village green. A landowner should therefore either make sure the land is inaccessible to the public, or erect notices making it clear that any access is on a permissive basis. There is no procedure to lodge a notice with the local authority to rebut any dedication of the land to the public as there is for public rights of way under section 31(6) of the Highways Act 1980.

Chapter 7
Planning

7.1 Introduction

Planning is surrounded by a considerable amount of law but the grant of planning permission itself relies on the judgement of the local planning authority (LPA) or the Secretary of State (usually acting through an inspector). Following the 2001 election, the Secretary of State is now the Secretary of State for Transport, Local Government and the Regions (DTLR); formerly, and from 1972, it was for The Environment (DoE), and since 1997 it was for The Environment, Transport and the Regions (DETR).

In any case, the Secretary of State's judgment should be made in the light of all material facts relating to the proposal, the policies in the development plan and the characteristics of the property itself. Landowners often wish to undertake works on their properties and a real tension can arise between the need to undertake such works on that owner's land and the policies of the LPA for the authority's area as a whole.

Under section 54A of the Town and Country Planning Act 1990 (the Planning Act), the LPA should determine any application for planning permission 'in accordance with the [development] plan unless material considerations indicate otherwise'. The LPA is also directed, under section 70, to 'have regard to the provisions of the development plan, so far as material to the application, and to any other material considerations'. The extent of such 'material considerations' is not defined; they embrace anything which has a bearing on the development proposal.

These statutory directions to LPAs are important. Regard must be had to the development plan and to the general advice from government published in circulars and policy planning guidance notes. Most planning applications, however, raise a significant number of other local issues, for example the safety and adequacy of access, noise, impact on landscape or the environment, or the impact on neighbours. Such matters should be considered before making any planning application so that the proposal can be presented to the LPA in a constructive way.

This chapter looks first at the impact of the planning rules on individual landowners and farmers; the last section deals with the development plan and other government guidance.

7.2 The need for planning permission

Planning permission is usually required for operational development which is defined as the carrying out of building, engineering, mining or other operations in, on, over or under land or the making of any material change of use of any buildings or other land (section 55(1) of the Planning Act).

Building operations are further defined so as to include the demolition of buildings, rebuilding, structural alterations of or additions to buildings, and other operations normally undertaken by a builder.

Demolition is a complex issue under the planning terms: partial demolition of any building is treated as a building operation and subject to the normal planing rules on the basis that the works result in a structural alteration to a building. Total demolition is different and written notice, with sufficient details of the works, must be given to the LPA of the intention to demolish any building so that the LPA can state if its approval is required. Under the Town and Country Planning (Demolition – Description of Buildings) Direction 1995 (Annex A to DoE Circular 10/95 *Planning Controls over Demolition*) certain buildings are excluded; planning permission is not required for the demolition of

(1) listed buildings, scheduled ancient monuments or buildings in conservation areas, because other controls apply to these;
(2) a building with a volume of less than $50\,m^3$, measured externally;
(3) any building other than a dwellinghouse or adjoining a dwellinghouse, and
(4) the whole or part of any gate, fence, wall or other means of enclosure unless it is in a conservation area.

Other activities are also expressly brought within the scope of development, namely

(1) the use as two or more separate dwellinghouses of any building previously used as a single dwellinghouse;
(2) the deposit of refuse or waste on land even if the land is already used for that purpose, if the area of the deposit is extended, or the height exceeds the level of the surrounding land;
(3) the removal of any material from a mineral working deposit;
(4) the placing or assembly of any tank in inland waters for the purpose of aquaculture; and
(5) the display of an advertisement on any part of a building which is not normally used for that purpose – but deemed consent for certain advertising is authorised under the Town and Country Planning (Control of Advertisement) Regulations 1992 (SI 1992/666).

Planning permission is not required for certain works and section 55(2) of the Planning Act includes

(1) the maintenance, improvement or alteration of any building where this only affects the interior or where this does not materially affect the exterior;
(2) the use of any buildings or other land within the curtilage of a

dwelling house for any purpose incidental to the enjoyment of the dwelling house as such;
(3) the use of any land for purposes of agriculture or forestry (including afforestation).

Planning permission is deemed to be granted for development which is authorised under the Town and Country Planning (General Permitted Development) Order 1995 (SI 1995/418) (GPDO).

The need to obtain planning permission is entirely separate from the need to secure consent under other legislation. Most importantly, the planning regime applied to listed buildings and land of special historic or environmental quality is significantly more onerous. Development opportunities are restricted in order to preserve the essential qualities of such properties and any breach of the rules usually involves the criminal law.

7.3 The General Permitted Development Order

7.3.1 Nature of the GPDO

This allows certain work to be undertaken without the need to obtain express planning permission, provided that it falls within the limitations of the appropriate class and the work complies with the relevant conditions. The GPDO designates 33 classes and many relate to specific types of development, including development by local authorities, drainage authorities, CCTV, aviation and minerals. The most important classes for the private landowner and occupier are described below.

Where development exceeds the specified limits, the whole of the development is regarded as unauthorised and enforcement action can be taken against the whole of the works.

7.3.2 The single private dwellinghouse

Part 1 of the GPDO permits a wide range of works to be undertaken without the need for express planning permission as follows:

(1) The enlargement, improvement or alteration to a dwellinghouse where
  (a) the cubic content of the original house is not increased by more than $115\,m^3$ (but less in some cases);
  (b) the works do not exceed the height of the original building;
  (c) the new works are not closer to the highway than 20 m or, if nearer, any part of the original building;
  (d) the works are not within 2 m of the boundary of the curtilage and more than 4 m high (save for works to alter or instal a window in an existing wall); and
  (e) the footprint of the buildings does not exceed 50% of the curtilage. Where a separate building exceeding $10\,m^3$ is erected

within 5 m of the dwelling house, it is treated as part of the dwelling for assessing the cubic content.
(2) Works affecting the roof of a dwellinghouse are permitted where the works result in an enlargement but the new roof does not exceed the original height nor project closer to the highway, and the increase in cubic content is kept within specified limits.
(3) The construction of a porch where the floor area is less than 3 m², the height is less than 3 m and the addition is not nearer than 2 m to a highway.
(4) The provision of any building or enclosure, including a swimming pool, where
 (a) it is no nearer the highway than the nearest part of the original dwelling or 20 m;
 (b) the height of the building does not exceed 3 m unless it has a ridged roof when it can be 4 m; and
 (c) the total footprint of the buildings does not exceed 50% of the curtilage.
(5) The provision of a hard surface within the curtilage.
(6) An oil tank not exceeding 3500 l, where the tank is not closer to the highway than 20 m or, if nearer, any part of the original building.
(7) One satellite dish subject to size constraints.

Three points in the above list are worthy of further examination.

7.3.2.1 Assessment of cubic content
The assessment of cubic content relates to the original dwellinghouse and the extent of any works authorised by the GPDO must have regard not only to the current proposals but also to the aggregate of all previous alterations and improvements. Cubic content is measured externally and relates either to the house as it was on 1 July 1948 or, if built later, as so built.

7.3.2.2 The enjoyment of the dwellinghouse as such
The second key point is that any works must relate to the enjoyment of the dwellinghouse *as such*, taking account of the requirements of a reasonable occupier, rather than the personal wishes of the current occupier. It has been held that the parking of a replica Spitfire aircraft, and the repair of stock cars or hang gliders are not covered by this rule. They require express planning consent.

The most common area of uncertainty is the use of a home for business purposes. Where the business can be carried on within the home without requiring any special alterations, planning permission is not usually required. Where works are undertaken expressly to create a working area or a specific office is created, planning permission should be obtained.

7.3.2.3 Within the curtilage

The final point relates to what is meant by 'curtilage' which is not defined in planning law. It is left for determination in each case, as a matter of fact and degree.

It is generally accepted that the curtilage is a small area of land forming part or parcel of a piece of land with the dwelling or building which stands on it, but the interpretation continues to be scrutinised by the courts. On examination in each case, the curtilage is identified as the area which reasonably and necessarily forms a single unit with the house and garden.

This does not mean that all land occupied as part of the garden or as amenity land with a dwellinghouse will fall within the curtilage. Nor is it necessary for the curtilage to be enclosed from other land. There are examples of the curtilage being defined so as to exclude areas of rough grass which have been mown and occupied by the residents of adjoining dwellinghouses.

The correct assessment of the curtilage can be important because development outside the curtilage requires express planning consent. In several cases tennis courts, swimming pools and other works have been held to be on land which does not form part of the curtilage. The LPA can require such works to be removed.

The curtilage of a listed building has a greater significance as the listing of a property covers all buildings within the curtilage. This was considered in the case of *Skerritt's of Nottingham* v. *Secretary of State* ([2000] 20 PELB). Double-glazed plastic windows had been installed in the stable block in the grounds of, but some distance from, a Grade II listed hall. It was held that the curtilage was not a small area but had to be identified by establishing the area which includes what are or have been ancillary buildings, in terms of ownership and function. The Court of Appeal held that the curtilage of the hall did include the stables.

7.3.3 Minor works

Part 2 of the GPDO also permits the undertaking of minor works. Of importance is the deemed planning permission for the erection or alteration of a gate, fence, wall or other means of enclosure up to 1 m high adjacent to a vehicular highway and 2 m elsewhere.

7.3.4 Changes of use

Part 3 of the GPDO links with the Town and Country Planning (Use Classes) Order 1987 (SI 1987/764, as amended). This distinguishes between various uses of properties and defines the changes that can be made to the use of land and buildings within each class and between some classes without the need to obtain express planning approval. For example, class B allows a change of use from storage and distribution (B8) or general industry (B2) to business use (B1) or from B1 or B2 to B8 provided the floor space concerned, in either case, does not exceed 235 m^3 (2500 ft^3).

Not all uses of land and buildings fall within the definitions set out in the Use Classes Order. These are called *sui generis* uses, requiring planning permission for any material change of use.

The enclosure of farmland as part of a garden is a material change of use and there is no exemption to cover this; express planning permission should be sought in every case. Where permission is given, it is likely to be subject to a condition which removes all permitted development rights to prevent the construction of outbuildings or other works.

7.3.5 Temporary buildings and uses

Two important exemptions are granted under part 4 of the GPDO: buildings and other works may be provided temporarily on any land to facilitate the undertaking of works on that land or on adjoining land. On completion of those works, the land must be restored to its previous condition. It is on this basis that contractors are exempt from the need for planning permission for compounds when building new roads or undertaking other works, and it is important to note that the exemption only applies to land 'adjoining' the works.

Land which is not within the curtilage of any building can also be used for up to 28 days a year for any purpose, except for a market/car boot sale and motor car/motorcycle racing when 14 days is the limit. This exemption does not authorise any building or engineering works and merely authorises the use of the surface.

The second exemption applies to the planning unit and does not permit an area to be subdivided in order to benefit from a number of such exemptions.

7.3.6 Caravans

It is an offence to keep caravans permanently on any land unless a licence has been obtained under the Caravan Sites and Control of Development Act 1960 and such a licence cannot be granted unless planning consent has been obtained.

Part 5 of the GPDO provides limited exemptions, including

(1) Use of a caravan within the curtilage, and for purposes incidental to the enjoyment, of the dwelling house
(2) Three caravans for up to 28 days in any year on holdings of over five acres
(3) Five caravans on recreational sites managed by exempt organisations, e.g. the Caravan Club and the Camping and Caravanning Club
(4) Meetings organised by exempt organisations, e.g. scouts and guides, lasting less than five days
(5) The stationing of caravan(s) on land incidental to the agricultural or forestry use of that land

In the case of farm and forestry businesses, consent is not required if a caravan is used only to accommodate seasonal workers or for the purposes of storage or shelter.

7.3.7 Agricultural buildings and operations

Provided that agricultural land is used for the purposes of agriculture and the development is reasonably required, part 6 of the GPDO permits the erection or alteration of any building other than a dwelling on holdings of over 5 ha where

(1) The work is not done on a separate part of the unit of less than 1 ha
(2) The building is designed for agricultural purposes
(3) The area involved does not exceed 465 m^2 (5000 ft^2)
(4) The height does not exceed 12 m, or 3 m within 3 km of an airfield
(5) No part of the work is within 25 m of the metalled part of a trunk road or classified highway
(6) The work of construction, enlargement, alteration or excavation is not undertaken within 400 m of the curtilage of a protected building and the purpose is to keep livestock or to store slurry or sewage sludge

A 'protected building' is any permanent building which is normally occupied, or built for occupation, by people; such occupation is not restricted to residential use. This does not include any building on the same holding or any dwelling or other building on another agricultural unit used in connection with agriculture.

Such a building can be used for livestock where there is no other suitable building and it is necessary to use the building for the purposes of quarantine, or because an emergency has arisen making other buildings unfit, or to provide temporary accommodation for animals normally kept outside. In any other case, express planning permission must be sought.

The right to undertake most works on an agricultural unit is subject to the notice requirement. An application must be made to the LPA so that it can decide whether its prior approval to the siting, design and external appearance of the building or works will be required. The application must be accompanied by a written description of the works, a plan showing the site and the requisite fee.

Development may not then begin until the LPA has said that its prior approval is not required, or the LPA has given its approval or 28 days have passed since the application was made without the LPA stating whether prior approval is required. Where prior approval is not required, the development must be undertaken in accordance with the application; where it is required, the works must comply with the approved details. This provision allows the LPA to discuss and seek changes to the proposals in appropriate circumstances but it cannot prevent the undertaking of necessary works unless an Article 4 Direction is applied (see section 7.3.9)

Where an agricultural building permanently ceases to be used for agricultural purposes within 10 years of its construction or significant extension or alteration (meaning that the cubic content of the original building is exceeded by more than 10% or the alteration means that the height of the original roof is exceeded), the LPA can require the building to be removed unless planning permission has been granted for its retention for non-agricultural use within three years of it ceasing to be used for agriculture.

To operate this provision, notice should be given to the LPA within seven days of the substantial completion of works to construct, extend or alter a building. The GPDO does not state what happens if such notice is not served.

More limited exemptions apply to works on units of less than 5 ha, the most important of which is that there is no right to erect a new agricultural building.

Two particular points arise in connection with part 6. The works must 'be reasonably necessary for the purposes of agriculture' and 'within the unit'.

The first point, '... reasonably necessary for the purposes of agriculture' allows the LPA to consider whether the works are properly related to the agricultural business. Recreational or hobby farming does not benefit from the rights. Even if the works fall within the size limits, works are not permitted where they are excessive given the nature of the business; buildings which are built of traditional materials and in a form which would be easy to convert to residential use can be challenged. Where the use cannot be justified, enforcement action can be taken.

The words '... within the unit' opens up an entirely different set of issues. The exemptions apply to works associated with the agricultural use of land and other buildings on the same unit. Given that many farmers now occupy land in different places and in different ownerships under a variety of agreements, this limitation can be significant. It has been held in a number of cases that a central storage facility on one unit cannot be used to store crops grown on other units – while it may be the case that the combined units form one agricultural business, they are separate planning units.

This does not cause a problem where a new building over 465 m^2 is to be erected as express planning permission must be obtained; it can result in the use of buildings erected under the GPDO being restricted, unless planning permission is obtained for a material change of use from agriculture to storage. This is even more the case if an agreement is made for one farmer to lease and use buildings belonging to a neighbour; and, in passing, it should be noted that rates can arise on such arrangements.

The agricultural exemption can be subject to further conditions, including in the case of the extraction of minerals for agricultural purposes, the material shall not be removed from the unit, and in the case of waste materials brought onto the site to construct a building or hardstanding, they must be used 'forthwith'.

The agricultural use of buildings is taken to include not only the storage of produce, but also its processing to prepare it for market and sale. This includes farm shops where all, or the substantial majority, of produce sold is produced on the holding. (This is different from the rules on rating under which the processing areas are exempt but the farm shop area is rateable.) It has also been held, in *Millington v. Secretary of State and Shrewsbury & Atcham BC* ([2000] JPL 297) that the processing of grapes into wine and its subsequent sale was a perfectly normal activity for a farmer growing wine grapes and therefore a use of land (and buildings) for agriculture. This does not mean that every processing activity will be treated as agricultural but where a crop is grown, processed and prepared for the market on the same holding, there is a strong presumption that the exemption will apply.

A further important issue is whether temporary buildings count as exempt or, by virtue of the works undertaken in their construction, as development requiring planning consent. Pig arks, polytunnels and bale buildings erected for lambing are usually regarded as exempt because they are not permanent and they are directly related to the use of the land for agriculture. That is not always the case. It has been held that the works involved in the erection (1) of polytunnels covering 2.6 ha and up to 4 m high, and (2) of a bale building with the support of scaffold poles both involved 'development'. In a second case involving Skerritt's (*Skerritt's of Nottingham v. Secretary of State* [2000] EGCS 43) it was held that a marquee erected adjacent to a hotel for eight months a year amounted to a building requiring planning permission because of its size, degree of permanence and degree of physical attachment to the land; it did not matter that the marquee was not present for 365 days in any year.

Where there is uncertainty about proposed works, there is a mechanism for establishing whether it requires the express grant of planning permission and this is described in 7.8.5 below.

7.3.18 Forestry buildings and operations

Part 7 grants exemptions for

(1) Works for the construction, alteration or enlargement of a building
(2) The formation, alteration or maintenance of a private way
(3) Operations on the same land or on other land held or occupied with that land to gain materials for (2) above
(4) Other operations for the purpose of forestry not including mining or engineering operations

The exemptions do not apply to works involving a dwellinghouse, buildings over 3 m high within 3 km of an airfield or within 25 m of the metalled part of a trunk or classified road. The same procedures for notification apply as for agriculture.

7.3.19 Article 4 Directions

LPAs have power to ban the exercise (1) of permitted development rights in any part of its area or (2) of some permitted development rights in all or part of its area. Most bans of these types can only last for six months unless confirmed or approved by the Secretary of State but an LPA can impose a permanent direction on its own authority where

(1) it concerns certain rights within conservation areas;
(2) it relates only to a listed building, a building notified to the Secretary of State as of architectural or historic interest or to development in the curtilage of a listed building; or
(3) it cancels any previous direction.

The DETR has given advice on the use of Article 4 Directions in Appendix D to Circular 9/95. The power should only be used in exceptional circumstances when there is a real threat that permitted development rights are likely to be exercised and that damage will be caused to an interest of acknowledged importance.

In particular, the circular states that when notified of agricultural or forestry works, an LPA should only issue a direction when there is a real and serious threat to amenity (and in other circumstances the 'amenity' of land has been judicially described as 'the look of the place'). It also refers specifically to the protection of land where its exceptional beauty or topography is particularly vulnerable to the indiscriminate exercise of permitted development rights.

7.3.10 Other restrictions

The exemptions are further restricted by the GPDO itself and by other legislation. The GPDO designates certain areas with limits on the freedom to exercise some of the above rights. Restrictions apply to prevent alterations or works which may adversely affect the character of the area without obtaining express planning approval. The primary areas of protection are National Parks, the Norfolk Broads, areas of outstanding natural beauty (AONB), sites of special scientific interest (SSSI), listed buildings and conservation areas. These are addressed in section 7.6 below.

7.3.11 Other authorised development

The Secretary of State can himself grant planning permission for development in certain circumstances. Government Orders, for example for railway works under the Transport and Works Act 1992, can also include the express grant of planning permission.

This power of the Secretary of State has been challenged in a number of cases on the grounds that the Secretary of State could not make a decision on a proposal where he was both the policy maker and the decision maker. The lead case concerned Alconbury airfield: it was decided in the High Court that the Secretary of State's dual role was in conflict with the right to a fair and impartial hearing under the Human Rights Act 1998 (*Alconbury Developments* v. *Secretary of State DETR* [2001] EGCS 5). The case was not heard in the Court of Appeal but went straight to the House of Lords where it was held that the planning process complies with the Human Rights Act: it was held that the Secretary of State could properly discharge both policy and administrative functions and that there is a safety net by way of the power of any person to seek a judicial review of any improper or unreasonable decision.

7.4 Obtaining planning permission

7.4.1 The application

In order to undertake development not covered above, an application must be made to the relevant LPA. In most cases this is the district or borough council but the county council is responsible for development involving minerals and waste. Where there is a unitary authority, it controls all planning matters.

Any application is governed by the Town and Country Planning (General Development Procedure) Order 1995 (SI 1995/419). The application should be made on the LPA's form and be accompanied by

(1) A plan showing the site and the proposal
(2) Details of the proposed works in adequate detail
(3) A certificate giving details of land ownership and any tenancy under the AHA 1986, and
(4) The appropriate fee

In addition, notice must be given in the prescribed form (supplied by the LPA) to any person other than the applicant who is an owner or an Agricultural Holdings Act tenant of the land.

The application may be made for outline planning permission or for detailed consent. Outline planning permission approves the principle of the development on the basis that details of any or all of the following matters will be submitted for subsequent approval:

- Siting
- Design
- External appearance
- Means of access
- Landscaping of the site

Where the LPA considers that it cannot determine an application in outline, it must advise the applicant within one month of registering the application, giving details of the additional information it requires. Outline permissions normally state that the reserved matters should be approved within three years and that the development should then be begun within two years of that approval.

A detailed planning permission allows the development to be undertaken in accordance with the application and supporting papers as submitted, subject to any amendments that may have been submitted to the LPA thereafter.

7.4.2 Fees

Planning fees are payable with most applications for planning consent. A fee is also payable with any notice served, for example, giving details about the erection of a new agricultural building. The amounts are prescribed under the Town and Country Planning (Fees for Applications and Deemed Applications)(Amendment) Regulations 1997 (SI 1997/37). The payment varies depending on the nature of the work and the site area, the number of houses or the floor area of commercial space, depending on the nature of the application.

7.4.3 *Planning conditions*

Section 70(1) of the Planning Act permits the LPA to grant permission unconditionally or subject to such conditions as it sees fit. For minor applications it may be that no conditions will be imposed but most schemes will be subject to a number of conditions.

Circular 11/95 puts this provision into context. The power to impose conditions is very wide and this can mean that permission can be granted for a development which would otherwise have to be refused. It is clearly stated that conditions must be

- Necessary
- Relevant to planning (and not some other purpose of the authority)
- Relevant to the development to be permitted
- Enforceable
- Precise
- Reasonable in all other respects

There are 79 model conditions set out in Appendix A to the circular for use by LPAs.

Where planning permission has been granted with an onerous condition, it can be challenged. This should not be done by way of an immediate appeal because this would allow the inspector not only to look at the condition but also at the planning permission – and he could overturn the original grant of planning permission. Instead, an application should be made under section 73 of the Planning Act to carry out the development without complying with the condition(s). If the LPA then refuses the application, an appeal can be made; the inspector will then only be able to decide whether the condition should or should not remain.

7.4.4 *Planning obligations*

These are different from planning conditions. Section 106 of the Planning Act allows a person with an interest in land which is the subject of a planning application to enter into an agreement with the LPA to

(1) restrict the development or use of land;
(2) require specified operations or activities to be undertaken in, on, over or under the land;
(3) require the land to be used in a specified way; or
(4) require one or more sums to be paid to the authority.

Obligations are usually entered into with the LPA. They can be made with other authorities in order to include other matters, for example a contribution to the provision of educational facilities by a county council or the adoption of open space by a parish council. Affordable housing requirements, for rent and for shared ownership schemes, are usually secured through section 106 obligations.

Obligations can be entered into unilaterally by the landowner: the signed deed can simply be handed to the LPA. This allows a person, who is pursuing a planning appeal against the refusal of an application, to make a binding commitment to do certain things if the appeal is successful, without having to secure the agreement of the LPA, which may be reluctant to discuss terms. The obligations are then enforceable by the LPA.

Obligations must again relate to the proposed development and they should be used to secure objectives which cannot be achieved through the use of conditions. Full details are set out in Circular 1/97 where it is stated that planning obligations should be

(1) Necessary
(2) Relevant to planning (author's note: i.e. not supposed to include other purposes of the authority, unrelated to land use planning)
(3) Directly related to the proposed development
(4) Fairly and reasonably related in scale and kind to the proposed development
(5) Reasonable in all other respects

Obligations should not repeat planning conditions nor deal with matters which should properly be dealt with by conditions.

Obligations should be drafted so that the provisions only bind the owner of the land for the time being; in this way a landowner can pass the liability to a purchaser. Further, the provisions should only bite when the planning permission has been granted and implemented; this allows an obligation to lapse with the associated planning permission if the development is not undertaken.

Where an option has been granted on land, the owner may be asked to enter into a planning obligation in order to secure a planning permission, on the basis that once the permission has been granted, the developer may exercise the option and buy the land. Agreements should provide that the owner need not sign the obligation unless the developer indemnifies him against the costs and/or has confirmed that he intends to buy the land.

The rules relating to such obligations changed on 9 November 1992, when Schedule 12 to the Planning and Compensation Act 1991 came into force. Obligations are different from planning agreements, which is the original term used to describe such arrangements, when entered into under section 52 of the Town and Country Planning Act 1971 or before 9 November 1992 under section 106 of the Planning Act, in its original form. Planning agreements can only be altered or discharged by agreement between the owner and the LPA or following an application to the Lands Tribunal for such alteration or discharge. In these proceedings, the agreement is treated as a restrictive covenant.

Planning obligations entered into after 9 November 1992 can also be altered or discharged by agreement, but in the absence of such agreement there is a process for their review. Five years after the agreement was entered into, any person against whom the obligation is enforceable may apply to the LPA for the obligation to be modified or discharged. If the application is refused, an appeal can be made to the Secretary of State.

7.4.5 Time limits for a permission

If it is not implemented, a detailed planning consent will last for five years unless a different period is specified. An outline consent lasts for three years within which details of the reserved matters should be submitted to the LPA and agreed; a further period of two years is then allowed for the development to be implemented, unless some other period is specified.

It is not usual for a planning permission to specify a completion date. Where the LPA is concerned that development may not be finished, it has the power to serve a completion notice on the owner and the occupier. That can require the work to be completed by a specified date, which cannot be earlier than 12 months from the date of the notice. A notice cannot take effect until it is confirmed by the Secretary of State and any person served with the notice has the right to be heard at an inquiry.

On occasions, it may be necessary to implement a planning permission in order to preserve the enhanced value of the land. This is particularly important now that the Secretary of State has advised LPAs, in policy planning guidance (PPG), to consider any application for the renewal of a planning permission on the same basis as a new application; the specific advice relates to housing in PPG 3, but this approach may be applied more widely. Given the role of the development plan and that planning policies can change, it cannot be assumed that all existing permissions will be renewed. It is prudent, therefore, to begin work, even if the project is then put on ice.

In the case of *Riordan Communications Ltd* v. *South Bucks DC* ([1999] EGCS 146) the court decided that work to implement a permission had started where

(1) the works were done in accordance with the permission;
(2) all reserved matters and details required by condition had been submitted to and approved by the LPA;
(3) the works were part of the authorised development;
(4) the works were not *de minimis*; but
(5) it is not necessary to intend that the works be taken through to completion.

There is a further caveat which means that if any other consent is required from the LPA, such consent must have been obtained. If the planning permission requires the details of any further matter (e.g. landscaping or drainage) to be submitted to the LPA and approved in writing, then such written approval must have been obtained.

7.5 Specific planning issues

7.5.1 Obtaining permission subject to an agricultural occupancy condition

The building of new dwellings in the countryside is strictly controlled. Planning permission can be given for the construction of isolated houses where it can be shown that the design is of the highest quality, and outstanding in terms of its architecture and landscape design, and that the house would significantly enhance the setting and wider surroundings. This is intended to continue the tradition of building country houses.

Otherwise, only those properties which are essential to allow farm or forestry workers to live near their place of work may be permitted, and these are subject to strict occupancy conditions. Annex I to PPG7 sets out the current policy guidance.

In order to justify a new agricultural dwelling five tests must be satisfied

(1) there must be a clearly established functional need;
(2) the need must be for someone wholly or primarily engaged in agriculture, and not part-time;
(3) the unit and agricultural activity must have been operating for at least three years, have been profitable for one of them and have a clear prospect of remaining profitable;
(4) there must be no other dwelling on the unit or elsewhere which is suitable and which could be used; and
(5) the proposal must satisfy the normal planning rules, for example on access and siting.

Of the above tests, those at (1) and (3) are primary tests to be considered first

(1) The *functional test* should establish whether it is essential for a worker to live at or near to the place of work in order to be on hand at most times, for example to provide essential care and supervision

of crops or livestock or to protect specialist crops from system failure
(2) The *financial test* should establish whether the agricultural business is economically viable.

The functional test is restrictive. Animal welfare is generally accepted as a proper basis for an application but concern about the security of a unit without animals is often not enough to justify approval. Also, there are conflicting decisions about the maintenance of technical systems serving high value crops, as remote sensors and warning devices have been considered adequate to alert a person some distance away.

The agricultural exemption does not apply to housing required by people involved in food processing, as compared with food production, but it is likely that the need for housing for other workers in an increasingly diverse rural economy is likely to be kept under review, but always under tight control. An agricultural dwelling should not be permitted where it is to be used for the housing of a retired farmer on his own land.

Agricultural dwellings are usually restricted in size. They should be of a size appropriate to their purpose of housing a worker, and they should not be unreasonably expensive to build. PPG7 does refer to the carrying cost of a new dwelling and some decisions have taken that into account. It should, however, be enough to show that the business can make a profit after funding any borrowings; if all or part of the cost is funded by the applicant personally, it should not be relevant that he chooses to waive any return on the capital, provided the business is viable.

LPAs also have the right to investigate whether the applicant has disposed of any property recently; if a suitable property has been sold, the LPA may consider that there is no justification for a new dwelling.

LPAs are aware that many owners want to secure permission for such a dwelling on their land, and that many such owners would rather have no occupancy restriction; accordingly, LPAs scrutinise any application with the greatest of care.

Conditions have been imposed on new farm dwellings over many years and they can be expressed in very different terms. It is important to consider the actual wording in every case as some may be unenforceable. Where a planning consent was granted for 'a farmhouse' but there was no occupancy condition, it was held that the permission did not restrict the occupancy. Where a condition restricted occupation to named people, it was held that once those people moved in, the condition would be permanently discharged on the grounds that a mere change in occupier would not be a material change of use.

When granted, the standard condition (paragraph I17, Annex I to PPG7) now reads

'The occupation of the dwelling shall be limited to a person solely or mainly working or last working in the locality in agriculture or in forestry or a widow or widower of such a person and to any other resident dependants'.

'Last working' does not mean that a person qualifies if he was working in agriculture and is now working in another business. It covers a person who is retired, ill and not working, or redundant.

Where an application is made for an additional dwelling on a unit, it is established that LPAs have the power to impose an occupancy condition not only on the new dwelling but on any existing dwellings. The existing properties are only bound by this restriction if the new planning consent is implemented. This can affect the value of the unit in the market and as security.

A further condition may be imposed on the new dwelling to restrict or remove permitted development rights under the GPDO to allow the LPA to control any extensions and outbuildings.

Comparable assessments apply to applications and permissions for forestry dwellings.

7.5.2 Removing an agricultural occupancy condition

If a dwelling is no longer required for business purposes, an application may be made to remove the condition. LPAs are reluctant to agree to this on the basis that they wish to retain a stock of dwellings for farm workers. If an application is to succeed, it must normally be shown that

(1) The property has been actively marketed by a reputable agent for at least six months, and sometimes longer
(2) It was advertised in the right journals and papers during the marketing period
(3) The price was reasonable
(4) No qualifying person in the locality is interested in the dwelling

The price should reflect the fact that the land does not have development value, as the permission was only granted for a farm business purpose. Typically this can result in a discount of 33% from the unencumbered market value, but regard should also be had to the cost of construction.

If it can be shown that a planning condition was improperly imposed, for example because there was no justification for the original grant or because the original application would have been approved in any event, the condition should be removed.

Where a property has been occupied for more than ten years in breach of an agricultural occupancy condition, it should be possible to secure a CLEUD (see section 7.8.5).

7.5.3 Horses

The use of land for grazing by horses does not always require planning permission unless it is associated with other horse activities. This is because the grazing of land does not constitute development under section

55(2)(e) of the Planning Act: it is a matter of establishing whether the horses are grazing the land primarily in order to be fed or whether they are being kept on that land for recreation or for exercise. Often this distinction is not pursued until a stable is erected.

Planning permission should be obtained for the erection of stables on any land unless the development is permitted by the GPDO within the curtilage of a dwelling house. Where stables have been erected without permission and have stood for more than four years without any action being taken or notice being served by the LPA, the stables are protected from enforcement action (see section 7.8.4).

Where a horse enterprise is run, for example as livery or a riding school, planning permission is required for the erection or change of use of any buildings and permission may be required for the land as well. There are conflicting opinions and, it will be necessary to consider the facts in each particular case.

7.5.4 Abandonment

It is possible for the permitted use of a property to be abandoned, even if this leaves the property with no 'planning use'. This most often arises in the context of dilapidated or derelict houses that have stood empty for many years. If an application is pursued to renovate and occupy such a property, the LPA must take account of four matters

(1) The physical condition of the building
(2) The time that has elapsed since it was last used as a residence
(3) Whether the building has since been used for any other purpose
(4) The intentions of the owner

Each of these criteria is to be given equal weight in assessing the application.

Planning permission can also be lost if permission is obtained to convert a building to an alternative use and it then turns out that the structure is such that the building has to be reconstructed. The rebuilding is treated as new development: where planning permission would not have been granted if the existing building were not there, the planning permission for the conversion is of no effect. The LPA can require the replacement building to be demolished.

7.6 Listed buildings, conservation areas and other designated property

7.6.1 More onerous rules

The rules relating to heritage property are significantly more onerous. Unauthorised development affecting a listed building, scheduled ancient monument or other property designated for its historic, architectural or environmental value under any part of the relevant legislation is usually a criminal offence. Unlike breaches of ordinary planning controls, enforcement action can be taken at any time.

7.6.2 Listed buildings

The Secretary of State has a duty to compile a list of buildings of special architectural or historic interest (section 1(1) of the Planning (Listed Buildings and Conservation Areas) Act 1990) (the Listed Buildings Act). To qualify, a building should

(1) be of architectural interest because of its design, decoration and craftsmanship or as an example of a particular building type or technique;
(2) be of importance in illustrating social, economic, cultural or military history;
(3) have close links with nationally important people or events; or
(4) have a group value in association with other properties, e.g. as a square or a model village.

There is no appeal against listing, but the Secretary of State can be asked to review the list. This process has no statutory basis but flows from the power for the Secretary of State to amend the list at any time. Also, if an application for planning permission and listed building consent for a development is refused and appealed, the appellant can then include the argument that the building is not of such a quality or character to merit listing within that appeal. The Secretary of State can then remove the property from the list if he considers that appropriate.

The listing of properties is normally undertaken on a programmed basis so that an area or a class of buildings is assessed, with the best being listed. Where a property is believed to be under threat from redevelopment or where new features have been discovered, the property can be 'spot listed'. This immediately imposes the full burden of the listed building regime and the need for listed building consent on the owner; this can frustrate development proposals when they are at an advanced stage. An application can be made to the Secretary of State for a certificate which will prevent the listing of the building for five years; an application for such a certificate invites a closer examination of the building and can result in it being listed anyway (section 6 of the Listed Buildings Act 1990).

As it is only the Secretary of State who can spot list a property, the LPA has the power to serve a 'building preservation notice'. Such a notice takes effect if fixed prominently on the building itself. The notice has the same effect as listing and it lasts for six months, during which time the LPA can ask the Secretary of State to add the building to the list. When a certificate is granted under section 6 a building preservation notice cannot be served for five years.

A listed building is defined so as to include the building itself, any object or structure which is fixed to the building and any object or structure which is within the curtilage of the building, forming part of the land since before 1 July 1948. This broad definition includes all buildings within the curtilage, and it can include such items as chandeliers in a house, and statues or ornaments in the grounds. To decide whether an object or structure is included, it is necessary to decide the degree and purpose of the annexation to the land or building.

While only buildings can be listed, the definition of any listed building includes its curtilage and the same rules apply.

Development is subject to strict controls in order to protect the character or setting of the building and listed building consent (LB consent) must be obtained.

If LB consent is granted for the demolition of a listed building, at least one month's notice must be given to the Royal Commission on the Historical Monuments of England (or the Royal Commission on Ancient and Historical Monuments in Wales) to allow the building to be recorded.

In deciding whether to grant LB consent, the LPA must consider whether the proposed development would affect the special interest or character of the building, the making of mere alterations is not enough to justify refusal as these may not affect the special interest or character.

If work is done without consent to a listed building, the LPA can serve an enforcement notice on the owner, the occupier and any other person with an interest in the property. If that notice is not appealed or it is upheld, the owner must comply. If he does not, the LPA can enter and undertake the works, recovering the cost from the owner. If the enforcement notice is appealed or an application is made for unauthorised works, LB consent can be granted for all or some of the works. The works covered by the LB consent then become authorised as from that date, but this does not prevent a prosecution being brought for the original undertaking of the works without authority, under section 9 of the Listed Buildings Act.

When the LPA is considering whether to take enforcement action in order to reverse unauthorised changes, it must consider whether it would have granted LB consent if an application had been made. If it would, enforcement action should not be taken.

There are rights of appeal in the event that LB consent is refused or if a listed building enforcement notice has been served. The procedures broadly follow the standard rules (see section 7.8 below).

If a listed building requires urgent repair, works can be undertaken without committing an offence where it can be shown that

(1) the works were necessary on grounds of safety, health or preservation of the building;
(2) this could not be achieved by providing temporary support;
(3) the minimum works were undertaken; and
(4) notice of the works and their justification were provided to the LPA as soon as reasonably possible.

The fact that the justification for unauthorised works is so tightly defined shows the serious attitude that is taken to the preservation of listed buildings.

7.6.3 Tree Preservation Orders

When an LPA is concerned about any individual trees, group of trees or an area of woodland which it considers important, it can pursue a Tree Preservation Order (TPO), under the Town and Country Planning (Trees) Regulations, SI 1999/1892. The TPO has to be made and copies are served on everyone with an interest in the land, giving the reasons why it has been made and a period of at least 28 days for objections. The LPA must consider the objections and then decide whether to confirm the order, with or without modifications.

An owner can seek consent to fell or carry out other work on a protected tree, group, or woodland. A refusal of consent can be appealed to the Secretary of State. If the owner is not allowed to undertake the works, he can claim compensation for any loss he may incur, but not if the loss

(1) is less than £500; or
(2) stems from the loss of development value; or
(3) was not foreseeable at the time of the refusal; or
(4) was foreseeable and the claimant has failed to take steps to avoid or reduce the loss; or
(5) is the cost of an appeal to the Secretary of State.

It is an offence to undertake works without such consent unless the tree(s) are exempt on the grounds that the tree(s) are dead, dying or in a dangerous condition, or because the work is necessary

(1) to comply with any obligations imposed by Parliament; or
(2) to abate a nuisance; or
(3) to implement a planning permission.

Other exemptions apply but they mainly flow from the grant of other express approvals.

Special rules apply to trees in conservation areas; for example, a tree may only be cut down if its diameter at 1.5 m above the ground is not more than 75 mm, or 100 mm where it is to be removed to improve the growth of other trees. Usually it is necessary to give prior notice of proposals to the LPA so that it can consider whether to make a TPO.

It is a criminal offence to undertake work on a protected tree without proper authority. Ignorance is no defence. Even where an independent contractor is engaged, the owner may be liable as it is an offence to cause or permit the works to be undertaken.

7.6.4 Conservation areas

LPAs have a duty to assess whether any parts of their area are of special architectural or historic merit which should be preserved or enhanced and any such area should be designated a conservation area. Although most conservation areas relate to built-up areas, there is no requirement for this to be the case and extensive areas of land, including parkland, can be designated.

The LPA must prepare a statement for each area to describe it, explain why it has been designated and to propose works to conserve and enhance the area. When considering any application for development within a conservation area or exercising any of its duties, the LPA must take into account the desirability of conserving or enhancing the character or appearance of the area.

The planning regime within conservation areas is more onerous and the following matters should be noted in particular:

(1) The demolition of buildings is controlled unless the whole building has a cubic capacity of less than $115\,m^3$
(2) Permitted development rights under the GPDO, are restricted: the limits on small extensions are tighter, exteriors cannot be clad, roofs may not be altered and satellite dishes may not be erected on chimneys or roofs, or on walls fronting a highway
(3) Trees are protected (see section 7.6.3)

At the time of writing, further proposals relating to the demolition of walls, fences, gates, means of enclosure, porches and chimneys have been published by DETR and further rules may be introduced shortly.

7.6.5 Other designated property

Substantial areas of land in England and Wales are designated, and for a variety of purposes. Special planning considerations apply in National Parks, in areas of outstanding natural beauty, to sites of special scientific interest, and to scheduled ancient monuments; these are all national designations imposed in recognition of the very special qualities of the areas.

Notwithstanding the GPDO, the Conservation (Natural Habitats, etc.) Regulations 1994 (SI 1994/2716) require express approval to be given under Regulation 62 before any work is undertaken if the works are likely to have a significant effect on a special area of conservation (SAC) or a special protection area (SPA), and the works are not directly concerned with, or necessary for, the management of the area.

The Town and Country Planning (Environmental Impact Assessment) (England and Wales) Regulations 1999 (SI 1999/293) also apply to the exercise of permitted development rights. Where the works fall within Schedule 1, an environmental study (ES) is required; where the works are in Schedule 2, the LPA can decide where an ES will be required. In general, an environmental impact assessment (EIA) will be required when it is likely to have a significant effect by virtue of factors such as its nature, size or location. Thresholds are included, below which it is unlikely that an EIA/ES will be required.

The 1988 Regulations included Schedule 2 developments where the works would be likely to have 'a significant effect on the environment'. In 1999, this was clarified and, in relation to agriculture, the rules apply to

(1) Projects for the use of uncultivated or semi-natural areas for intensive agriculture where the area exceeds 0.5 ha
(2) Water management projects, including drainage and irrigation, exceeding 1 ha
(3) Intensive livestock installations exceeding 500 m^2
(4) Intensive aquaculture installations designed to produce more than 10 tonnes of dead weight fish per year
(5) Any reclamation of land from the sea

Importantly, the above regulations apply, without regard to the above thresholds, where the relevant land is designated as a sensitive area, being

(a) An SSSI
(b) Land within 2 km of an SSSI where English Nature or the Countryside Council for Wales has notified the land as 'sensitive' to the LPA
(c) Land subject to a Nature Conservation Order
(d) Within a National Park or the Norfolk Broads
(e) Land designated as a property on the World Heritage List
(f) A scheduled ancient monument
(g) Within an AONB
(h) Within an SAC or SPA

7.7 Appeals

7.7.1 Procedure

The LPA should determine every planning application within eight weeks but some can be delayed because the LPA requests more information or because the proposal is complicated or contentious. The applicant has the right of appeal to the Secretary of State at any time after the eight week period, unless the applicant and the LPA agree to extend the time allowed for the decision to be made; there is no right of appeal six months after the eight week (or extended) period expires. It is sensible to record any agreement to extend the period for decision in writing.

When an appeal is submitted, the LPA no longer has the power to determine the application. Where the LPA does not decide an application for a long time, the applicant may hope that continued discussions and adjustments may result in a favourable decision, but time limits, for example in a contract for the sale or lease of land, may require an appeal to be made. For this reason, a second application may be submitted for the same development at the outset or at a later date, to allow one of the applications to be appealed within the time limits, while the LPA continues to deal with the parallel application. This can be an expensive option as each application must be accompanied by the full fee.

Since 1 August 2000, where a planning application is refused, the LPA must specify all policies and proposals in the development plan which are relevant to the decision. The applicant may then appeal to the Secretary of State under section 78 of the TCPA 1990. The Planning Inspectorate will act on behalf of the Secretary of State; the inspector will usually be appointed to hear and decide the appeal but in some cases, of greater significance, the inspector may report to the Secretary of State for him to make the final decision.

Planning appeals are made on the form obtained from the Planning Inspectorate, and a copy should be sent to the LPA. The appeal must be accompanied by any of the following documents which are relevant to the appeal

(1) The application with all plans, drawings and documents
(2) All correspondence with the LPA relating to the application
(3) Any certificate relating to owners or tenants of the site
(4) Any other plans, drawings or documents relating to the application but which were not sent to the LPA
(5) The decision notice
(6) If the appeal relates to the approval of matters relating to another planning permission, then the application and approval of that other permission

If any of the above documents are relevant and are not received within the six-month period, the appeal can be rejected outright.

Appeals can be decided by written representations, allowing the inspector to make a decision after reading the submissions of the appellant and the LPA. Each will have the opportunity to see what the other has submitted and to comment upon it. The inspector will visit the site and then decide the matter. The inspector will make an accompanied site visit to allow particular features to be pointed out, but if the site can be seen readily from the public highway an unaccompanied visit will be offered.

Appellants can, however, ask for an inquiry so that the arguments can be put to the inspector in person. Each side will then have the opportunity to cross-examine the representatives of the other side, the effect being to subject the arguments to closer, adversarial scrutiny. This process is subject to a strict timetable for the provision of information and the submission of papers. The appellant and LPA are also required to prepare a statement of common ground four weeks before the inquiry.

The Inspectorate may offer the parties the opportunity of an informal hearing: this aims to explore the issues in a more informal way, with the inspector identifying the key matters and leading the discussion.

7.7.2 *Judicial review*
Any person who believes that a planning decision has not been properly

made, whether granted by the LPA or allowed or rejected on appeal, may seek a judicial review. The application must be made promptly and certainly within three months of an LPA decision, but within six weeks of an appeal decision. It is this safety net which has been endorsed by the House of Lords in the context of the law on human rights. In order to secure a hearing, it is necessary for the applicant to show that he has a proper reason for challenging the decision, in that he must be affected by it, and that the decision has been made erroneously.

It can be argued that the LPA or the Secretary of State has acted improperly and that the decision is unlawful. It is also possible to argue that the decision has been made without regard to important elements of the development plan or policy guidance issued by the government, and that if such guidance had been taken into account, a different decision would have been reached. In such cases, the court may agree that the decision is unlawful: the application will then be returned to the LPA or the Secretary of State for the matter to be considered anew.

The court will not substitute its own judgement for that of the LPA or the Secretary of State where the decision has been lawfully and properly made, even if the court feels that the decision was wrong. LPAs and the Secretary of State have considerable flexibility in the exercise of their discretionary judgement in deciding whether to grant or refuse an application or appeal. The courts interfere in that process with great reluctance.

7.7.3 Third party rights of appeal

At present, the LPA notifies near neighbours of a planning application and, for some proposals, notices are put up on site and details are published in the local paper. In this way people can know about and comment on the proposals before the decision is made.

There has always been a concern that the grant of planning permission can have an adverse effect on other owners and occupiers. Government has considered how this should be dealt with so that those who are materially affected by a scheme can be sure that the decision is taken properly, in the light of all relevant facts, and with proper regard for everyone's concerns.

Government believes that further, and unacceptable, delays and problems would be likely to arise if third parties were given a right of appeal to the Secretary of State against the grant of planning permission by an LPA. Third party rights of appeal are not to be introduced but this does not prevent any such person, who can show that they have a proper interest in a planning decision, from seeking judicial review.

It remains to be seen what effect the introduction of the Human Rights Act 1998, from 1 October 2000, will have on the attitude of the courts to the interests of third parties where their home is adversely affected by the grant or implementation of a planning permission; it should be noted, however, that there is no right to a view in English law. It is unlikely that a third party will be able to claim that his possessions have been interfered with when planning permission has been granted so that his view has been affected. This, however, remains to be seen.

7.8 Enforcement of planning control

7.8.1 Powers
The planning system fails completely unless there are effective mechanisms to ensure that unlawful activities can be controlled effectively. LPAs have been given a number of powers to achieve this under the Town and Country Planning Act (1990, as amended). All statutory references in this section are to the Act unless otherwise stated.

7.8.2 Planning contravention notices
Where development has taken place without the grant of deemed or express planning permission, the works are unlawful. An enforcement notice can be served on the owner and occupier specifying the development which has occurred in breach of the rules and the steps that the owner and/or occupier should take to remedy the matter.

It may not be clear to the LPA what is being done on some land and by whom. The LPA can serve a planning contravention notice on anyone who is the owner or occupier of the land in question or on any other person who is carrying out activities on that land. The recipient must provide the information requested so long as it concerns

- Any use of the land or operations or other activities carried out on the land
- Any conditions or limitations in a planning permission granted on that land

The notice may also ask specific questions about

(1) Suspected uses, operations or activities
(2) The dates when such uses, operations or activities began
(3) The people who have been involved
(4) Any planning permission or why this is not required
(5) The nature of the recipient's interest and for him to supply details of anyone else known to have an interest in the land

7.8.3 Breach of condition notices
Where the breach of control involves a failure to comply with a planning condition, the LPA can serve a notice (under section 197A) on the person with control of the land requiring compliance. The notice must specify the steps that should be taken to comply and a reasonable period within which to achieve this.

There is no appeal against this form of notice except for the recipient to show that he took all reasonable steps to secure compliance or that he no longer has control of the land.

7.8.4 Enforcement notices

This is the basic weapon for an LPA faced with the unauthorised development or use of land. An enforcement notice may be served (section 172) on an owner or occupier by the LPA if it is expedient to do so, having regard to the development plan and all other material considerations. This approach clearly shows that the breach of planning control is not itself an offence; it is the failure to act upon an enforcement notice which creates the offence.

An enforcement notice must be served on the owner and occupier of land and on any other person whose interest in the land would be materially affected. The notice must specify

(1) the matters alleged to constitute a breach of planning control;
(2) the steps to be taken to remedy that breach; and
(3) the period for taking such steps – this cannot be a period of less than 28 days from the date of service, and different periods can be specified for different steps

An appeal may be submitted against an enforcement notice (section 174). The effect of this is to suspend the notice until the appeal has been determined. An appeal, which will be subject to strict time limits in line with the normal procedures, may be made on any or all of the following grounds:

(1) Planning permission ought to be granted, or the condition or limitation should be discharged
(2) The alleged activities have not occurred
(3) The activities are not in breach of planning control
(4) On the date of the enforcement notice the activities were exempt from enforcement
(5) Copies of the notice were not properly served
(6) The steps required to be taken are excessive
(7) The time allowed for taking those steps is insufficient

On an appeal, the Secretary of State may grant planning permission for any or all of the development, vary or discharge any planning condition in a planning permission, uphold or quash the enforcement notice, and/or correct any defect in the notice where this does not cause injustice to any party.

Once an enforcement notice takes effect, by being upheld on appeal or if no appeal is made, the notice must be complied with. Failure to comply is an offence punishable by a fine up to £20 000. If the matter is taken to court, the failure to comply with a notice is construed strictly: the court will not entertain any review of the planning arguments, as these should have been settled at an inquiry on appeal, whether or not an appeal was made.

The LPA has the power to enter the land to undertake the works specified in the notice and recover the cost from the owner.

Where the LPA is concerned that activities will continue and that it is expedient to prevent this, a stop notice can be served (section 183); this can be sent with an enforcement notice or served at any time until that notice takes effect. A stop notice must specify a date not less than 3 or more than 28 days from service for compliance.

Compensation can be payable as a result of a stop notice if the enforcement notice is quashed or varied so that the activities which have been prevented are accepted as authorised or lawful, or if the stop notice itself is withdrawn.

Unauthorised development cannot be the subject of successful enforcement action, by enforcement notice, stop notice or breach of condition notice, when it can be shown that the breach has existed for more than either four or ten years (section 171B), on the following basis:

(1) The four year rule applies to all operational development undertaken without planning permission. The period is measured from the date when the works were substantially completed. It also applies to the change of use of any building or part of a building to a single private dwellinghouse; this has been held to include conversions to holiday cottages where each cottage has the same facilities as a dwellinghouse.
(2) The ten year rule applies to all other development carried out without planning permission. It includes any material change in the use of land, the carrying out of development without complying with one or more of the planning conditions, and to any breach of a condition or limitation; the last point includes any continuing breach of an occupancy condition imposed on a dwelling.

7.8.5 Certificates of Lawful Development

Where an activity is protected from enforcement under the four and ten year rules, a Certificate of the Lawfulness of Established Use or Development (CLEUD) can be obtained (section 191). An application should be made to the LPA describing the use or development which has been carried out on the land. Evidence should also be submitted to show the extent of the activities and when they first started, together with any supporting documentation to show that the breach has been continuous thereafter.

The LPA will look closely at any application but it is obliged to determine the application by considering the evidence submitted and any other relevant information available to it. The certificate must be issued if the LPA is satisfied on the balance of probability that the activity has been carried on as stated in the application for the relevant period. The LPA does not have to be absolutely certain that the activity has been carried on, but it must establish that this is likely, on the balance of probability.

There is a right of appeal to the Secretary of State if a certificate is refused. It is also the case that the refusal of a certificate does not prove that the development has not in fact been carried out for the requisite period; it means that the information provided has not been proved sufficient. If more evidence can be found, a new application may be made.

If it turns out that any of the information submitted with the application was false in a material way or that material information was withheld, the LPA can revoke the certificate. No compensation is payable on such revocation. The LPA should give notice of an intention to revoke a certificate as this could cause serious problems for the owner, especially if the land has been bought at a value to reflect the certificate; an owner should have the opportunity to make representations before a revocation is made.

If a CLEUD is sought for a dwelling on the grounds that it has been occupied in breach of an occupancy condition for more than 10 years, the evidence must be very clear. If the same person has occupied the property throughout the period, that person should remain in occupation at the time when the application is made and when it is determined. If a person moves out and the property is left empty for a period before being reoccupied, the breach will probably be treated as having ceased when the first occupier moved out; a new breach will begin when the next person moves in. The position is unclear when one non-qualifying occupier moves out and another moves in; given that the change from one occupier to another is not a material change of use, the breach is likely to be treated as continuous if there is only a short period when the property is left empty.

Where a person proposes to use or develop land and is unsure whether planning consent is required, a Certificate of Lawfulness of Proposed Use or Development (CLOPUD) can be sought (section 192). An application can be made to the LPA with details of the proposal, asking it to decide whether the proposed use or development would be lawful.

Once issued, a CLOPUD is conclusive proof that the development is lawful, provided that there has been no material change in circumstances between the date of issue and the start of the use or works, such as an order removing relevant permitted development rights on the land.

7.9 The development plan

It is unusual for the development plan and government guidance (see section 7.10) to be left until last in a summary of planning law. This is the case here because, although development plans are at the heart of the planning system and government planning guidance is important, they are a mixed blessing to those who seek to promote a planning application.

The development plan for any area comprises two parts: in most rural areas the county council prepares the *structure plan* to set out the strategy for its area, and each district or borough council prepares a *local plan*, allocating sites for larger developments and setting the framework for the assessment of other projects. (In unitary authorities, the two parts are kept separate but published by the same Council as part 1 and part 2 plans.)

The real problems arise because the development plans take a very long time to prepare and they are subsequently updated from time to time. The hierarchy can be summarised.

(1) The government issues *policy planning guidance notes* (PPGs) on a number of subjects. These explain the issues which are important in the formulation of any proposal and how LPAs should assess them.

(2) The county councils and unitary authorities, which are grouped into Regions, meet to discuss the major strategic issues for their area, including housing requirements, employment, transport, water issues, etc. Each regional meeting will produce a suggested form of *regional planning guidance* (RPG) for consideration by the Secretary of State and for public comment. At the end of the process in each region, the Secretary of State will issue regional planning guidance.

(3) The County Council will then prepare its *structure plan*. This is a strategy document setting out policies to deal with issues in broad terms, for example allocating the required housing and employment requirements between the districts and setting out framework policies for all other matters. This plan is published for public comment, with comments being required in a specified six week period; all comments must be considered by the council. There is also a requirement for the Council to have an examination in public (EIP) to allow the opportunity for wider debate. There is no right to be heard at this inquiry, it is by invitation. EIP does not consider all issues in the draft plan, but focuses on the major and contentious issues. Unless the Secretary of State calls in the structure plan for his own consideration, amendment and publication, the council can agree and adopt the structure plan.

(4) The district and borough councils then prepare the *local plan*. This contains considerable detail. It can show sites allocated for development and draw lines on plans, for example to define the boundary of an *area of high landscape* or to show development boundaries in order to contain the majority of new development within appropriate sites. The local plan will also include policies setting out the LPA's approach to applications on unallocated sites, for example redevelopment or expansion of large premises in the countryside or the replacement of a dwelling in open countryside. A *deposit draft local plan* is published for comment; when representations have been received, the LPA should meet objectors and discuss their comments, in order to try to resolve the matter. Thereafter, and subject to any amendments which have been made, a second deposit draft local plan is published. Further objections must be made in another defined six week period and any objections which are made and not

withdrawn, must be considered at an inquiry. The objections can be dealt with wholly in writing or be considered in person, with the giving of evidence, and the examination and cross-examination of witnesses. Usually an inquiry is held.

When the inquiry is finished, the inspector will report to the council. It must publish that report within eight weeks of receipt. The LPA does not have to accept the inspector's recommendations, but if the LPA then decides to make changes to the deposit plan it must publish those changes and allow a further period for comments and objections. If those comments cannot be resolved satisfactorily another inquiry must be held. This cumbersome system allows people to propose sites for the LPA to allocate for development in addition to, or substitution for, other sites in the deposit plan. If the inspector agrees, the LPA cannot include those sites without giving other people the opportunity to assess the impact of that inclusion, and to object to the proposal if they so choose. Only at the end of that process when the council's proposals have been approved after public consideration can the local plan be adopted. This process does again raise issues under the Human Rights Act, as the council acts both as policy maker and decision maker, but it is too early to comment on the changes that may be made to address this.

(5) *Mineral and waste plans* are different again. They are treated as local plans setting out detailed policies and making site-specific allocations. They are prepared by the county council which is the mineral and waste planning authority. The process for making objections and holding an inquiry is the same as for other local plans.

(6) The effect of this process can be confusing because each RPG, or structure or local plan is for a defined period and each needs to be kept under review. The process is intended to be sequential so that each plan is based on the plans which have been agreed at a higher level, but it takes a long time to carry each of the plans through from the assessments at the start of the process to final adoption. This means that the local plan for a period may be prepared and discussed while the original structure plan is itself being reviewed, with proposed changes to strategic policies, because the RPG has been published in a new form.

Under section 54A of the Planning Act, the LPA should determine the application in accordance with the development plan unless material considerations indicate otherwise. The structure plan as approved and the local plan as adopted by the LPA form the 'development plan'. Emerging plans and guidance are treated as 'material considerations' and regard can be had to them. When the plans are out of sequence and the policies in the local plan conflict with policies in a replacement RPG or structure plan before those documents have been published or approved, uncertainty prevails. Certainly such lacunae can present opportunities but, more importantly, they complicate the assessment of individual applications.

Where an LPA proposes to grant permission for development which does not comply with the development plan, it must give notice to the Secretary of State, who has the power to call in the application and determine the matter. This is intended to bolster the development planning system but the Secretary of State only calls in about 100 such applications in any year. Applications are called in when they raise matters of more than local concern, either in the regional or national context because of the scale of the proposals or their impact on an interest of acknowledged importance (e.g. the environment or designated heritage sites). Otherwise, the Secretary of State is prepared to allow LPAs to exercise their discretionary judgement and make decisions even where they do not accord with the development plan.

Such a system is important for owners and farmers. The development plan process cannot be ignored when a large development scheme is to be promoted. When submitting any planning application, regard must be had to the relevant policies. No-one can, however, rely on the certainty of the development plan process. It is for this reason that the need for planning permission and the process for obtaining it were put first in this chapter.

7.10 Government guidance

In addition to regional planning guidance, government publishes policy planning guidance notes (PPGs) and mineral planning guidance notes (MPGs). These cover a wide range of topics and are essential reading for anyone promoting a development scheme. It is not appropriate to summarise them but a list of the PPGs and MPGs can be seen on the internet at http://www.planning.dtlr.gov.uk/policy.htm

Government also publishes circulars to explain how it interprets the law and how it wishes the LPAs to apply that law. These circulars do not overrule the strict interpretation of the law but they are extremely useful in understanding how the law should be applied. They provide advice on the way in which an LPA should carry out its duties in a reasonable way, balancing the interests of the individual against the interests of the wider community as a whole. A list of the circulars can be seen at http://www.planning.dtlr.gov.uk/advice.htm

Chapter 8
Compulsory Purchase and Compensation

8.1 Introduction

Most landowners resent the use of compulsory purchase order (CPO) powers to deprive them of their property but such powers have existed for many years and government has made it clear that they will remain to facilitate projects which are of general benefit. Government recognises that it is a serious matter to deprive people of their property, and that their rights must be respected; this approach is now enshrined in the Human Rights Act 1998. A CPO should only be pursued where the acquiring authority can show that it has a compelling case in the public interest. The procedures described below are intended to ensure that this balance is maintained.

Most landowners and occupiers do not believe that the system is fair because those who seek CPOs have resources and expertise available to them which few private individuals can match. Private owners also believe that it is unreasonable at best for them to lose all or part of their property, although government states that the risk of a CPO should be treated as another of the risks borne by any owner of property.

Statutory powers of compulsory purchase have existed for many years, to enable canals, railways and other works to be undertaken and the powers have been used by private undertakings and by government and other authorities; in a recent review of the principle of compulsory purchase government and all political parties have accepted the need for these powers to remain. They are considered to play an essential part in securing necessary development for public benefit.

8.2 The legislative background

The powers to acquire land were brought together in the Land Clauses Consolidation Act 1845 so that it would no longer be necessary to restate the powers in every new Act authorising the acquisition of land. Since then there have been numerous Acts and the majority of the present law is set out in

- The Land Compensation Act (LCA) 1961, which contains the main rules for assessing compensation for land taken
- The Compulsory Purchase Act (CPA) 1965, addressing matters both of procedure and additional compensation

- The Land Compensation Act 1973 (LCA), extending the right to compensation to other owners, including rights to compensation for some who do not have land taken and introducing home and farm loss payments
- The Acquisition of Land Act 1981 (ALA), which sets out the procedure for the granting of CPOs in most cases
- The Town and Country Planning Act 1990 (TCPA), which governs all development proposals and includes rights for planning authorities to use CPOs for proper planning purposes

In 1991 the Planning and Compensation Act made many further changes to the above Acts, affecting both procedure and compensation, and introducing blight notices.

The process which an authority must follow to secure a CPO is not straightforward, but it is relatively well explained in leaflets and guidance to any person whose land is affected. The real complexities emerge in the assessment of compensation when land is taken. It is not enough to interpret the words of the statute because many of the current rules have been laid down by the courts.

While many of the judicial decisions are clear, some are not. In one case in 1974, Lord Wilberforce commented on the history of section 68 of the 1845 Act (now incorporated into CPA 1965, at section 10(2)) as follows:

> '[section 68] has over 100 years received through a number of judicial decisions, ... an interpretation which fixes upon it a meaning having little perceptible relation to the words used. This represents a century of judicial effort to keep the primitive wording – which itself has an earlier history – in some sort of accord with the realities of the industrial age.'

In *Wagstaff v. Secretary of State DETR* [1999] 21 EG 137, a claimant who had no land taken sought compensation for the effects of the building of a new road. In a preliminary hearing at the Lands Tribunal, it was held that a right to compensation did arise but only in relation to the construction of one part of the scheme; the other parts of the scheme were to be ignored completely. It was said that the assessment of compensation on this basis, taking account of only a part of the works, might 'cause certain conceptual problems' but that 'such problems are not infrequently encountered in the law of compensation'.

These two comments make it clear that the assessment of compensation can be complicated; it requires a detailed knowledge of the rules and a mental agility to rationalise and assess values in circumstances which may not arise in the real world. For this reason, this chapter can only lift the corner of the veil for those involved in complicated cases.

8.3 The power to acquire land compulsorily

8.3.1 Procedures

A CPO can only be made where Parliament has given its authority. It can permit the acquisition of land or rights in land, and the interpretation of 'land' is very wide: it includes any buildings or structures, any interest or right which already exists in or over land, and any freehold or leasehold interest. In some cases, a CPO may create new rights in or over land which would be created for the scheme, but these can only be pursued if the statutory power permits such new rights to be created.

Where part only of a house (or park or garden belonging to a house), a building or a manufactory is to be acquired, the owner can serve notice to require the authority to take the whole of the property. If the acquirer objects, it can refer the matter to the Lands Tribunal for determination. The Lands Tribunal can decide whether part only can be taken without material detriment to the owner; or, in the case of a park or garden, without seriously affecting the amenity of the house.

Special procedures apply to any CPO which includes (1) land held inalienably by the National Trust, and (2) operational land owned by any statutory undertaker and used in the performance of its obligations: the specific authority of Parliament is required in these cases before a CPO can be confirmed. Where allotments are involved, the minister is required to certify either that alternative land is being made available or that this is not required in certain circumstances.

In most cases the powers are granted for specific purposes, but in general terms (e.g. for highway works or for the proper planning of the area). The land to be taken for any particular scheme is not defined; this allows the authorised body to prepare a scheme, identify the necessary land and pursue a CPO in order to undertake the works. Powers are included in many Acts and the detailed rules are set out below.

In some cases, for example the construction of the Channel Tunnel Rail Link, there is a special authorising act, which includes details of all the land that may be acquired. All objections to the project or any acquisition are considered during the passage of the bill through Parliament: because the bill affects individuals, this process is called the *hybrid bill* procedure. This approach is now unusual.

In order to reduce the number of hybrid bills which had to be dealt with, the Transport and Works Act 1992 was brought in; it lays down an alternative procedure for the development of railways, tramways, inland waterways and other transport systems, where a series of different ownerships must be acquired. A draft order must be published by the promoter, setting out the details of the works, any CPO powers being sought and the boundaries of all land to be acquired permanently or occupied during the construction period.

In pursuing a CPO, an authority must

(1) establish that the necessary power has been granted by Parliament and proceed in pursuit of that particular power: if the power does not exist or the wrong power is used, the CPO cannot be granted;
(2) secure planning permission for the scheme so that it can be shown that, if permitted, the CPO can be exercised; and
(3) have the intent at the time of making the CPO and when seeking its confirmation to implement the scheme: it is not however a requirement that the scheme is built following the purchase (but see section 8.5 on the Critchel Down rules).

In most cases, a CPO is prepared and 'published' by the authority with the statutory powers. The Secretary of State has the power to make CPOs himself; in such cases he publishes a *draft order*.

In either case, the acquiring body must identify all necessary land and set out details of all owners and occupiers in a book of reference. When the CPO is published, it must be advertised and notice must be served on all the owners and tenants (but not tenants for a month or less). The scheme must be available for inspection and a period of not less than 21 days must be allowed for the making of objections.

Once a CPO is made, land may be excluded but no land can be added to it unless all owners and occupiers agree.

8.3.2 *The inquiry*

If objections are made by any statutory objector, and are not withdrawn, an inquiry must be held. A statutory objector is any owner, lessee or occupier of the CPO land, except for tenants of a month or less; this includes any tenant or occupier of a house under an agreement which had an initial period of a month or less, and which has continued on that basis.

An inquiry allows the scheme to be considered in detail and for the effect of the CPO on the owner or occupier of property to be compared with the benefit of the scheme. The arguments presented at an inquiry can be very wide-ranging and are admissible so long as they are relevant except in the following two cases.

(1) They relate to matters of compensation alone: these are excluded because disputes over compensation are decided by the Lands Tribunal if the CPO is confirmed and compensation cannot be agreed.
(2) They relate to planning issues which have already been clearly established. If the land required for a scheme has been included in an adopted local development plan and the plan has been the subject of a local plan inquiry, the CPO inspector will usually not allow the planning arguments to be considered in the CPO inquiry. If, however, the precise site has not been identified and the adopted local plan proposes a scheme in general terms, the planning arguments and the merits of alternative sites can be considered.

Inquiries are governed by strict procedures. These are set out in the Compulsory Purchase by Non-Ministerial Acquiring Authorities (Inquiries Procedure) Rules 1990 (SI 1990/512) and in Compulsory Purchase by Ministers (Inquiries Procedure) Rules 1994 (SI 1994/3264).

The Regulations prescribe the conduct and timetable for each inquiry. These include the requirement for the acquiring authority to prepare a statement of case with full details and justification for the CPO. Statutory objectors may also be asked to serve a statement of case but it is usual for the objection to be included in proofs of evidence which must be exchanged between the parties in advance of the inquiry.

An objector may represent himself at the inquiry or may employ advisers, including counsel, to represent him. Each party must pay his own costs, but the authority should repay the reasonable costs to an objector if the CPO is refused. If the CPO is modified because of an objection, the objector will usually be given a partial award of costs and the authority must pay for the costs incurred in presenting that part of the evidence which was accepted.

After the inquiry has been held, the inspector will report to the Secretary of State and set out his recommendations, or explain why he is not making any recommendations.

If the minister decides that he disagrees with his inspector on any matter of fact or takes into account any new evidence or matter, he cannot determine the CPO until he has notified the parties, with details, and considered their comments. At the time of writing (February 2002), this ability for the Secretary of State to differ from the inspector, when he is also the minister responsible for administering the CPO process, raises concerns about the independence of the inquiry. This may be a breach of Article 6 of the Human Rights Act 1998 – the right to an independent and impartial hearing.

When a CPO is confirmed by the Secretary of State, it may only be challenged on the grounds (1) that the statutory authority for the CPO does not exist or (2) that the rules have been breached. A challenge can only be made within six weeks of the date when the confirmation of the CPO is first published and, if such a challenge is not made, the CPO is valid in all circumstances.

8.3.3 Exercising a CPO

After a CPO has been confirmed, the acquiring body has three years in which to serve a *notice to treat*, and a further three years in which to serve a *notice of entry*, giving not less than 14 days notice, and thereafter may take possession of the land. As an alternative the authority may make a *general vesting declaration*, which has the effect of transferring all interests in the land to the acquirer on a specified date; this approach is more usual when there are numerous interests to be acquired.

The notice to treat must be served within three years of the date of confirming the CPO. Thereafter, the acquirer must either (1) agree the compensation, or (2) refer the matter to the Lands Tribunal, or (3) take entry within three years. If none of these occurs, the notice to treat ceases to have any effect and the CPO can no longer be enforced. These periods mean that an owner or occupier can be held in suspense for many years, not knowing whether a CPO will in fact be implemented.

8.3.4 The blight notice procedure

When a *blight notice* is served, the authority is required to purchase the owner's interest unless an appeal is made and upheld. Paragraphs 21 and 22 of Schedule 13 to TCPA 1990 state that a blight notice can be served for

(1) land authorised to be acquired by a private Act or within the limits of deviation within such an Act; or
(2) land within a CPO (including the acquisition of rights in land) where the CPO has been confirmed but a notice to treat has not been served, or where the CPO has been published by the acquiring authority (or in draft by the Minister) but has not been confirmed, once notice has been served on owners and occupiers under paragraph 3(1)(a) of Schedule 1 to the Acquisition of Land Act 1981.

Other paragraphs in Schedule 13 may be of assistance to owners in other circumstances.

A blight notice can be served by the owner-occupier, which includes (1) a tenant where the lease has not less than three years to run and (2) a mortgagee in limited circumstances, of

(1) a house;
(2) an agricultural unit; or
(3) other premises with a rateable value of less than £24 600.

The claimant must have been in occupation for six months up to the date of serving the notice, although exceptions may be made where the occupier moved out in the 12 months before service and the property has been empty thereafter.

The authority can either accept the blight notice and purchase the property or serve a counternotice on one or more of the grounds set out in section 151 of TCPA 1990. A farmer can serve a blight notice to cover both the land within the CPO and the remainder of the unit, where the remainder cannot reasonably be farmed on its own or with any other land he may farm. In serving a counternotice, the authority can seek to restrict the CPO to the land which is actually required, and argue that the remainder can be farmed on its own. After the inquiry which will follow the contested blight notice, the blight notice may (1) be held to be valid and apply to all of the land, (2) be restricted to the area covered by the CPO, or (3) if the authority has satisfied the Lands Tribunal on one or more of the grounds for opposing a notice, be held invalid.

If the blight notice is accepted on all the land, the valuation of the land required for the scheme follows the normal rules. The additional land is valued on the basis of its current use, and no account is taken of any development or hope value.

All other owners and occupiers must wait until the CPO has been confirmed and notice to treat has been served to know if they will in fact be dispossessed. A notice to treat is equal to an exchange of contracts and either party can thereafter press for the compensation to be settled and paid.

There can be a benefit in the delay in that the owner of any interest can sell, gift or otherwise dispose of his interest throughout the process: the position changes only after the notice to treat, after which the acquirer is not liable for any increase in the compensation arising from the owner's actions. The ability to arrange one's affairs in this way can be important, given that the amount payable in compensation can, in some circumstances, be reduced to reflect the increase in value of other land which has not been taken (see section 8.4 below).

8.3.5 Missing owners

Where the owner of any land cannot be identified, special rules apply. The land can be taken and the acquirer must pay a sum assessed by a valuer member of the Lands Tribunal into court, the sum being the proper compensation for the land. The money will be held by the court and may be claimed by any person who can show title, and if dissatisfied by the amount he may refer the matter to the Lands Tribunal. Any money which has not been claimed after 12 years can be recovered by the authority.

8.3.6 Purchases by 'agreement'

Where an authority seeks to buy land by agreement, even though compulsory powers exist, there are some important issues to be borne in mind.

(1) The purchase will be in the conventional manner, and the contract and transfer will set out the full agreement between the parties. There will be no further right to claim compensation for matters which may have been overlooked.
(2) If payments for disturbance and other consequential losses are to be claimed at a later date, the details must be included in the contract.
(3) The authority cannot accept conditions when it has the power to acquire and use land without such conditions. This means that a restrictive covenant will be ineffective, even if it is included in the documents.

8.4 The assessment of compensation

8.4.1 Entitlement to compensation

In simple terms, when a CPO is exercised, the owner of land is entitled to compensation for

- the market value of the land taken;
- any resulting loss to the value of adjoining land which is retained through 'severance', being the loss of the land taken, or 'injurious affection', being the execution and use of the works on the land taken; and
- any other costs arising directly out of the acquisition.

The owner is not able to claim compensation for

- any increase in value generated solely as a result of the scheme;
- any loss which is unrelated to the acquisition under the CPO; or
- any other loss which may arise but which is 'too remote' from the acquisition.

Many problems flow from this approach: the underlying principle is that any person who has land acquired compulsorily should be paid sufficient money to put him in the same position as he was before the acquisition, in so far as money can achieve this. The effect of this is to limit the payment to the amount of the loss which can be identified and quantified, even though the owner of land usually feels that the losses arising out of the CPO are much greater.

It is important to note that there can only be one claim in relation to any scheme (unless no land is taken and claims can be made for the execution of works under CPA 1965 and for the use of works under LCA 1973 (see sections 8.4.14 and 8.4.15 below). Once a claim has been settled, there is no right to go back and raise other issues.

8.4.2 Compensation for land taken

The basic principles for assessing compensation are laid down in section 5 of LCA 1961 and compensation for land taken is paid on the following basis:

(1) no allowance is made for the fact that the acquisition is compulsory;
(2) the value is the amount which the land would be expected to realise if sold by a willing seller in the open market;
(3) the special suitability or adaptability of the land for any purpose is ignored if it could only be so used under statutory powers or bought for that purpose by an authority with CPO powers;
(4) any use which is illegal or a danger to health is ignored;
(5) where the land is put to a special purpose for which there is no market, the compensation may be assessed on the basis of equivalent reinstatement; and
(6) rule (2) (above) does not affect the right to compensation for disturbance or other matters not related to the value of the land.

The principle of market value is set out in rule (2) (above) but there are several parts to this: the owner is to be treated on the basis that he is willing, but he is not assumed to be 'anxious'. He is prudent and assumed to offer the land properly in the market in a way which will attract all potential bids; on the other hand, the acquiring authority is not regarded as 'willing' because this would imply that it might have to pay a price which reflects the value of the land to the scheme; the acquirer is deemed to match the best bid that would have been made.

This raises a significant issue: rules (2) and (3) above taken together provide that where the special value only arises out of the scheme and the scheme can only be taken forward by a body with CPO powers, the value will take no account of the special value. The compensation will remain at the level which could otherwise have been achieved in the market.

Rule (3) above, however, was amended in PCA 1991 so that full account can be taken of the value of the land to any other party who would be interested in acquiring the land in the market. It may be that land taken under a CPO has a general value in the market at one level but a premium value to one 'special purchaser', e.g. a neighbour. The rules now allow the full value which could be expected to be paid by that special purchaser to be taken into account, including the price to which he would go to secure the land. It is not to be assumed that the special purchaser would buy at a slightly higher level than everyone else who may be in the market (the 'one more bid approach') when the value to him is significantly higher.

The potential of the land to a special purchaser can create significant value – and argument. Major opportunities arise when the land could accommodate an access for the development of other land, when that development could not otherwise happen – the 'ransom' situation. This issue was first addressed in the case of *Stokes* v. *Cambridge City Council* ([1961] 13 P&CR 77), where the value of Mr Stokes's land was reduced by 30% to reflect the price that he would have had to pay for access to develop his land, if the CPO had not arisen. The 30% was considered appropriate in the circumstances of this case; the same principles arose in other cases (e.g *Hertfordshire County Council* v. *Ozanne* ([1989] 2 EGLR 18) and *Batchelor* v. *Kent County Council* ([1990] 1 EGLR 32)) but each has to be assessed on its own facts. While the principle of ransom has been approved in the courts, the assessment of compensation in a case by the Lands Tribunal does not establish any binding precedent.

If there is a special value to a third party, the amount and the probability of the third party making a bid for the property can be assessed by considering the role of a speculator; this is an approach which the Lands Tribunal has adopted on some occasions. The value in the market would reflect the amount that a speculator would bid for the land in order to profit by selling the land on to the third party: if the third party could bid for other land unaffected by the scheme or if there were only a slight prospect that the third party would buy the land, the enhanced value would be modest; if there is a high probability, the speculator's theoretical bid would be much higher.

Conversely, not all land which is suitable for a special purpose has a market value: in the case of a new sewage works to serve a village the value to the statutory undertaker is ignored as no-one other than the authority would be prepared to instal a sewage works.

8.4.3 The planning assumptions

In many cases, the value of land is based on its existing use, but planning permission is assumed to exist for the authorised works. Also, claims can take account of any development for which deemed or actual planning consent exists and for any other purpose for which planning consent would be granted. If land is used for a purpose but there is no express planning permission, the use is not held to be illegal if it is immune from enforcement action, under the four or ten-year rules (see section 7.8.4). To the extent that the market would reflect it, hope value can be included.

The fact that planning permission has been granted for the scheme is taken into account, but this is of no value if the permitted use is one which has no value to anyone other than the acquiring authority.

There is a provision (under section 17 of LCA 1961) for claimants to obtain a *certificate of appropriate alternative development*, in order to establish all uses for which the land would have been given planning consent in the absence of the CPO. An application should be made to the LPA for a certificate by letter and, while it is usual to suggest the uses which should be approved, this is not necessary: the LPA should set out every use for which it would have granted planning consent on the valuation date.

This matter has become more complicated in the light of local development plans: these set out the development policies for each district in general or specific terms. If land is identified within the plan for the authority's scheme, it will not be identified for other uses in addition. As the LPA should determine planning applications in accordance with the development plan, this can affect the ability to obtain a certificate for any other development. The granting of a certificate stating that no development would be permitted can reduce the hope value that might otherwise have been claimed.

Where the land is subject to an agricultural tenancy, the landlord cannot use the existence of planning permission for the authority's scheme to serve notice on the tenant in order to claim vacant possession and higher compensation.

Where land is granted planning permission within 10 years of purchase and the compensation would have been greater if that permission had existed on the valuation date, the former owner is entitled to the payment of additional compensation. The additional value is assessed at the original valuation date, not at the date of the permission; statutory interest is payable on the amount.

8.4.4 Betterment and the Pointe Gourde *principle*

Rule (3) in section 8.4.2 excludes the value of the scheme itself from the assessment. This was clarified in the case of *Pointe Gourde Quarrying and Transport Company Ltd* v. *Sub-Intendent of Crown Lands* ([1947] AC 565) where a quarry was acquired to provide limestone for the building of a naval base on another site. It was accepted that the profits from the quarry would be increased as a result of the supply of stone to the naval base but this was excluded from the compensation on the basis that the value arose solely because of the authority's scheme. If the naval base was not being built, such profits would not be made.

The essence of this rule is that any effect on the value of the acquired land arising from the public purpose(s) prompting the acquisition, whether from the adoption of the powers by the authority or from their implementation, is to be disregarded.

Finally on the market value basis for assessing compensation, the value of the land taken has to be assessed in isolation. No account is taken of any adjoining land which is also being acquired; in the 'no-scheme world' the claimant could not assume that the neighbouring land would also be available to any purchaser. The consequence of these provisions is to limit compensation to the value, over and above the land's market value for its current use, if any, which the owner could have achieved in the market in the absence of the scheme and the CPO.

8.4.5 Equivalent reinstatement

Compensation is rarely paid on the basis of equivalent reinstatement, under rule (5) in section 8.4.2, and four matters have to be considered

(1) the land must be used for a particular purpose and would have continued to be so used but for the CPO;
(2) there must be no general market for that purpose, and this is established if it can be shown that the owner could not expect to sell the property to any other person for that purpose at a fair price;
(3) there must be a genuine intention to use the compensation to acquire and use other property for the same purpose; and
(4) there is no legal right to compensation under rule (5) in section 8.4.2, and the Lands Tribunal has discretion whether to allow payment on this basis.

8.4.6 The valuation date

The date on which the value is to be assessed is defined, not in any statute but through case law. The date is

(1) where the values are agreed, the date of agreement;
(2) where the values are not agreed but the authority takes possession of the land, it is the date of entry; and
(3) in any other case where the matter is referred to the Lands Tribunal, it is the date of the award (being the last day of the hearing); but
(4) where compensation is awarded under rule (5) (in section 8.4.2), the

date is that on which the claimants can reasonably begin work to replace that which has been taken.

8.4.7 Settling the claim

The claim is submitted by the owner to the authority. This can be done when replying to the notice to treat which asks for details about title to the land being taken. There is an obligation to respond to the notice to treat but this does not extend to the claim for compensation. When a claim is submitted it should specify what is being claimed for and the amount for each item.

If a claim is not submitted within 21 days (section 4, LCA 1961), the acquirer can refer the matter to the Lands Tribunal. This may also affect the Lands Tribunal's decision on the owner's right to an award of costs.

In the absence of a detailed claim, the authority may withdraw its notice to treat and not pursue the CPO.

Once a claim has been made, the owner and the authority should seek to settle it. Either party may refer the matter to the Lands Tribunal at any time, but under the Limitation Act this must be done within six years of the valuation date. If an application is made for leave to refer the matter after that time, the applicant must show that there is good reason why the six year rule should be ignored; if leave is not given, the right to compensation is forfeit.

If the claim is referred to the Lands Tribunal, it is resolved in a quasi-judicial fashion. Strict rules apply to the procedure. Significant costs can be incurred. An owner may be awarded some or all of his costs, depending on the outcome of the hearing; this depends on any offers to settle made by either party in the run-up to the hearing and how these compare with the Lands Tribunal's award. In any event, when costs are examined in detail, or subjected to legal taxation, the amount paid rarely covers the bills; two-thirds is probably the most that is likely to be recovered.

8.4.8 Advance payments

Once the authority has taken possession, the owner is entitled to an advance payment. The request must be made in writing and the payment is due within three months. If the request is made more than three months before entry, the payment is due at the time of entry. This is 90% of the compensation as agreed or, if there is no agreement, 90% of the proper estimate of the compensation assessed by the authority's valuer. Interest is payable on all unpaid compensation from the date of entry at the statutory rate. As negotiations proceed, if the level of compensation increases, additional advance payments can be requested. If, when the compensation is settled, the advance payment is found to be greater than the sum due, the claimant must refund the excess.

8.4.9 Interest

Simple interest is payable on compensation from the date of entry. The rate is 0.5% below the base lending rate of the seven largest authorised banks in the UK, with special provisions if these banks do not all charge the same rate.

Interest is taxed in the year of receipt. If substantial sums are involved, interest should be paid on an annual basis where that amount exceeds £1000.

8.4.10 Severance and injurious affection

In many cases, particularly on farms and estates, where only part of a property is taken, compensation is payable for the reduction in value of any adjoining land caused by the taking of the CPO land, which is 'severance', or for the depreciation caused to the retained land by the execution and use of the CPO works, 'injurious affection', (section 7 of CPA 1965).

In assessing compensation for severance no account is taken of the acquisition of land from third parties; compensation is paid for the loss in value of retained land because of the loss of other land from the same ownership. Compensation for injurious affection, however, takes account of the execution of all the works and their subsequent use.

The basis of compensation is usually assessed by comparing the value of the property before the exercise of the CPO and the value of the remaining land on completion. Part of this difference will represent the value of the land taken and the remainder, the diminution in value of the retained land.

A claim for severance cannot be made, however, if the claim for the land taken is based on its development or other value, where that value could only be secured in the market by giving up possession anyway.

It is not necessary for the owner to hold the same interest in the land taken and the land retained. It is enough, for example, for the owner of land to have an option to acquire land included in the CPO for a claim for injurious affection to arise on other land which he owns or occupies.

It should also be noted that while there is no general 'right to a view' under the law, compensation can be claimed for this when the view is affected and the value of retained land is reduced.

8.4.11 Betterment

Where retained land increases in value as a direct result of the scheme for which the other land was taken, the increase in value is taken into account in assessing compensation (section 7, LCA 1961). The betterment must be directly related to the scheme and two scenarios illustrate this as follows:

(1) Where land is acquired for a new road and the same owner secures consent on adjoining land for a petrol filling station to serve drivers on the new road, the increase in value for the filling station can be taken into account; if the increase in value exceeds the compensation, there is no basis for the additional benefit to be recovered by the authority. The compensation may simply be nil. If, on the other

hand, the construction of the new road removes a constraint on other land so that it could be developed with access from another highway, that would not be treated as a direct result of the scheme and there would be no set-off for betterment.

(2) Where a contractor uses an owner's retained land as a depot while building a scheme or for the disposal of spoil, there will be no set-off unless the arrangements (a) are made so that they are binding on any successor in title to the land, and (b) are entered into before the valuation date. If the arrangements are not binding or are concluded after the valuation date, only the hope value reflecting the prospect of such an agreement can be set-off; given the options available to contractors in most cases, the hope value is usually accepted as nil.

8.4.12 Compensation for disturbance when land is taken

Rule 6 of section 5 to LCA 1961 clearly states that compensation based on the open market value does not affect the right for any person to claim disturbance; this is intended to cover the costs necessarily incurred by the occupier as a consequence of being displaced. Again, the claim for disturbance does not arise if the compensation for the land taken is based on a value which could only be realised if the owner gave up possession.

Horn v. *Sunderland Corporation* ([1941] 2 KB 26 established the principle that an owner is entitled to compensation which is the higher of

(1) the value of the land taken for a purpose which would require possession to be given if sold in the open market; and
(2) the existing use value of the land plus the costs incurred in relocating from it.

It may be that the capital or rental cost of new premises is higher than that for the property taken, but this is not taken into account. It is assumed that a person will receive 'value for money' when making any investment and that if the cost of new premises is higher or they require alteration, there will be a commensurate benefit from the investment.

Disturbance compensation is assessed on the basis of the cost to the claimant and there are three tests to be applied:

(1) the cost must not be too remote from the acquisition of the property and it must be the reasonable and natural consequence of the acquisition;
(2) the loss must have arisen as direct consequence of the acquisition; and
(3) the owner must have taken all reasonable steps to minimise the cost.

There can be serious problems when seeking to minimise the disruption of a scheme, particularly on a commercial business, because disturbance claims can only be made when a notice to treat is served. When a CPO has been confirmed, however, a prudent owner will take steps to protect his business, and probably begin a search for new premises. If successful and the business is moved before the notice to treat is served, compensation should be paid for disturbance but if the move could have reasonably been achieved at lower cost at a different time, the compensation may be restricted. If the notice to treat is not served, there will be no right to claim compensation.

This matter was illustrated in two cases: in *Prassad* v. *Wolverhampton Borough Council* ([1983] JPL 449) it was held that fees incurred in the search for a replacement property before the service of the notice to treat were allowable. In the Hong Kong case *Director of Land and Buildings* v. *Shun Fung Ironworks Ltd* ([1995] 2 WLR 404) it was held that compensation was payable for lost profits where the company could not enter into long-term contracts with customers on account of the proposed acquisition.

8.4.13 Accommodation works

There is no statutory right to accommodation works but they are often undertaken when part only of a property is taken to address the two following elements of compensation:

(1) The loss of part of a property can cause problems as a new boundary will be created and it will need enclosing, and a water pipe or an access may be disrupted. The accommodation works are provided to overcome the disturbance caused to the occupation of the land.
(2) Other works may be undertaken by the authority. The cost of these is usually compared with the compensation which would be payable in their absence. With some latitude, the authority will seek to settle on the option which costs least.

Two points are worth noting: accommodation works can include the cost of replacement boundaries, including new stone walls, or the construction of long access roads beside a highway; such works then become the responsibility of the owners and there can be real concern about the future replacement cost. While claims may be submitted for the replacement in 20 or more years based on the amortised cost, compensation is still based on market value; the future burden of these works (which can be significant) will only be reflected to the extent that it affects the market value on the valuation date.

There is a well established principle about the provision of new stock fencing, but the legal basis is obscure. The landowner or occupier is normally responsible for all stock fencing after completion of any scheme, but this does not apply where the boundary adjoins a railway or a motorway. For railways, the law is laid out in the Railway Clauses Consolidation Act 1845; for motorways, this obligation arises out of the highway legislation, which makes the Highways Agency responsible for the exclusion of certain hazards from motorways, including stock.

8.4.14 Compensation for disturbance when land is not taken

There is a separate provision which allows compensation to be claimed for the loss arising out of the construction of works (section 10, CPA 1965). The provisions are very restrictive and are called the *McCarthy rules*, following the case of *Metropolitan Board of Works* v. *McCarthy* ([1974] LR 7 HL 243). The rules state

(1) the loss must have been authorised by Parliament;
(2) if no such authority existed, there would have been a right of action in law;
(3) the damage must arise from an interference with some right in land which adds value to the claimant's property; and
(4) the damage must arise from the execution of the works, not their use.

The valuation date is normally the date the works are completed; interest runs on the unpaid compensation from the date of claim, which should be submitted as soon as the loss can be assessed accurately.

The House of Lords put a new interpretation of these provisions in *Wildtree Hotels Ltd* v. *Harrow LBC* ([2000] EGCS 80). The claim was for loss of profits to the hotel caused by restrictions on access, the obstruction of roads, and problems from noise, dust and vibration during the execution of the works. Compensation was payable but only to the extent that the various disruptions affected the underlying value of the property; the law does not provide for payment of the lost profits themselves where no land is taken.

8.4.15 Compensation for the use of public works

Part 1 of LCA 1973 permits compensation to be claimed for the loss in value of property caused by the use of public works, following their construction or alteration. Public works are defined as

(1) Any highway
(2) Any aerodrome
(3) Any other works on land undertaken in the exercise of statutory powers

The third category does not distinguish between public authorities or private sector undertakings, provided that statutory powers are exercised. Claims can only be made if the authority has immunity from an action in nuisance.

Claims can be made if the loss exceeds £50. They must be based on the loss caused by (1) noise, vibration, smell, fumes, smoke, or artificial lighting, and/or (2) the discharge of any solid or liquid substance onto the claimant's land.

If the view from a property is affected by artificial lighting, even at a distance, a claim may be made. Other factors, including any additional effect on the view from the presence of the works, are ignored.

The valuation date is the first anniversary of the date on which the works are first brought into use. No account is taken of any additions to the property which were first occupied after the claim date, even if they were under construction; no account is taken of any potential value for the actual or potential grant of planning permission for any change of use.

The right to make a claim runs with the property, even if it is sold after the works have been constructed. If there is a sale before the claim date, the vendor can reserve the compensation: he should make a claim after the exchange of contracts but before completing the sale (or granting a lease) and he will then be paid compensation based on the difference in value of the property in the condition it is on the date of completion.

If the property is not occupied when the claim is made, there is no right to compensation. The owner must either be in occupation or have let the property.

Claims can be made on the valuation date by

(1) the occupier of a house where he is the owner or the holder of a lease with not less than three years remaining on the date of claim;
(2) the owner-occupier of an agricultural unit; or
(3) the owner-occupier of business premises where the rateable value is less than £24 600.

As only one claim can be made in relation to any public works, full account must be taken of the future use of the works, including forecasts for the growth of traffic or flights.

The claim must also relate to the use of the works; where a highway is altered those who suffer loss as a direct result of the use of that altered section, may claim compensation; those who suffer from an increase in traffic but on existing, unaltered highways have no such right.

Interest is payable on the compensation from the later of the valuation date or the date on which the claim is made.

8.4.16 *Miscellaneous matters*

The right to acquire land compulsorily can relate to the acquisition of rights in land rather than the land itself, and such rights can be created anew for the purpose. This issue is considered in Chapter 9, Utilities.

In most cases when a home is taken, a *home loss payment* will be paid to the freeholder, or a tenant where the lease has not less than three years to run, provided that the person has been in occupation for at least 12 months and treated the dwelling as his or her only or main residence. The payment is 10% of the value of the home, with a minimum of £1500 and a maximum of £15 000. For occupiers who (1) have a statutory tenancy, (2) occupy under a restricted contract or (3) have a right to occupy under the terms of employment, the payment is £1500.

Where the whole of a farm unit is taken a *farm loss payment* can be claimed when replacement land is acquired. The payment is made to reflect the fact that the profitability will be reduced in the first few years while the farmer gets to know the land. The basis for calculating the payment is complicated, the payments are not very large and few claims have been made.

On the acquisition of any land the *minerals* are excluded unless they are expressly included in the CPO. Instead the *Minerals Code* applies and this has the effect of reserving the minerals to the original owner. If permission can be secured for the working of the adjoining minerals such that the minerals under the land taken would have been worked, then a further claim can be made at that time. For more information on this aspect, refer to Chapter 9, Utilities.

Following the exercise of a CPO, or blight notice, the owner is entitled to have his reasonable costs reimbursed. These cover two heads:

(1) Solicitor's expenses in advising and concluding the transfer of the land; these are paid on an hourly basis, with power for the costs to be subject to taxation.
(2) Surveyor's costs for negotiating and settling the compensation; these are normally paid on the basis of a standard scale (Ryde's Scale), calculated by reference to the amount of compensation, including the value of any accommodation works negotiated by the surveyor, and the nature of the property and interest being acquired.

8.5 The Critchel Down rules

These rules provide that land which has been compulsorily acquired should be offered back to the former owner. They are often quoted but have only resulted in the re-acquisition of land in a few cases. In a study published by the DETR in July 2000, information was available on 22 000 disposals of land by authorities since the introduction of the rules in 1954: the rules were held to apply to 3000 cases, the land was offered back in fewer than 200 cases, and only 26 former owners repurchased their property.

The Rules, as now set out in a DoE letter dated 30 October 1992 (not issued as a Circular), require any government land to be offered back to its former owners or tenants when it has not materially changed in character and

(1) the land was agricultural land and bought after 1 January 1935, or if bought after 30 October 1992 it was bought less than 25 years before the date it becomes surplus; or
(2) non-agricultural land which becomes surplus less than 25 years after its purchase.

The following further restrictions apply to remove the obligation to offer the land back where

(1) the ministerial authority (a) confirms that the land is needed by another department or (b) that it is appropriate for the land to be disposed of quickly to a local authority or other body with CPO powers;
(2) small areas of agricultural land would have no satisfactory agricultural use, even when used with other land of the former owner;
(3) it would be mutually advantageous for boundaries to be adjusted with an exchange of land;
(4) the land has been acquired from more than one owner and and is to be sold for development or where it comprises land where part only has changed in character, and there is a risk that a fragmented sale will not achieve the best price; or
(5) the market value is so uncertain that the best price cannot be assured on a private sale.

The freeholder has the right to repurchase unless the sale relates to a tenanted house, in which case the tenant has the first option, or in the case of other property, it was let under a long lease and there is a significant period unexpired at the date of disposal.

The rules apply to all government bodies, and to all land bought with compulsory powers or by agreement, where compulsory powers existed at the time. The rules are commended to, but not binding, on private sector bodies with compulsory powers.

Chapter 9
Utilities – Rights and Wayleaves

9.1 Introduction

Many organisations have a duty to provide essential services, including water and electricity, to homes and businesses; others provide supplies which are desirable if not necessary, for example cable television. Most suppliers are companies in the private sector but operating within a strictly regulated environment in order to ensure that a reasonable level of service is provided and at reasonable cost.

The utility companies provide water, sewerage, electricity, telecommunications and gas to properties. Other companies move petrol, oil and gas in bulk. Each of these services has the benefit of rights granted by Parliament to allow it to meet its obligations, including

- The right to pursue a compulsory purchase order to acquire land for a necessary purpose
- The right to seek an order to acquire rights in land, including the creation of new rights, rather than acquiring the freehold of the land itself
- In the case of the water and sewerage companies, the right to serve notice and enter land for the purpose of pipe-laying works
- In the case of electricity and telecommunication companies, the right to seek a wayleave for the placing of cables in or over land under an agreement for a year or a term of years

The acquisition of land itself is subject to the normal rules for compulsory purchase, which are covered in Chapter 8. This chapter considers the acquisition of rights in land, either permanently or under a wayleave. It is important to note that each of the statutes contains express provisions for the relevant service, and no two statutes grant precisely the same powers or duties.

9.2 Wayleaves

Even though this term has been used, and such rights have existed, for many years, there remains a legal uncertainty about what a *wayleave* is precisely in law. The author believes that it can be put in context to assist those who have to deal with the matter on the ground.

(1) An *easement* is a right in *servient* land which benefits *dominant* land. The owner of the dominant land is able to enforce his rights against the owner of the servient land. A private right of way is an example.

(2) The right to acquire a right in land compulsorily differs from an easement because the right imposes a burden on the landowner but for a specific purpose unrelated to any dominant land. The right, once taken, however, then exists on an equivalent basis and any transaction is subject to the existence of the right.

(3) A *wayleave* is a temporary right granted for a period of years or on an annual basis, which the owner of the land can terminate in certain circumstances. Wayleaves may be sought compulsorily under the Telecommunications Act 1984 and the Electricity Act 1989; each Act lays down different rules for obtaining and paying for such rights.

9.3 Works in the highway land and private tracks

Where rights are acquired in private land compensation is payable, but this is not the case when rights are exercised in highway land, and sometimes in private tracks. Public highways are treated by statute as a route not only for people and traffic but also for other services provided for public benefit. While the subsoil of many highways is in the same ownership as the adjoining land, Parliament has expressly provided that the subsoil can be used to carry services without charge. Further, some utility companies have the power to operate in private land as of right, and in some circumstances without payment. This can be explained for the main services by reference to statute and the following summary reveals clearly how the position can differ.

9.3.1 *The standard definition of a street*
Utility companies have power to use the subsoil in roads and tracks where they form part of a 'street', subject to any limitation in another Act. A 'street' is defined by section 48(1) of the New Roads and Streets Works Act 1991 (NRSWA) as

'the whole or any part of the following, irrespective of whether it is a thoroughfare –
(a) any highway, road, lane, footway, alley or passage;
(b) any square or court; and
(c) any land out as a way whether it is for the time being formed as a way or not.

Where a street passes over a bridge or through a tunnel, references ... to the street include that bridge or tunnel.'

9.3.2 *Other relevant legislation*

9.3.2.1 *The Telecommunications Act 1984*
This Act applies to all licensed telecommunication businesses, including cable TV. It gives companies the power to instal (which includes keeping,

maintaining, repairing and replacing) apparatus in a street which is maintainable at the public expense, but this does not apply to footpaths and bridleways which cross, or form part of, any agricultural land or land which is being brought into use for agriculture (paragraph 1 Schedule 2). Where apparatus is to be installed in a right of way in agricultural land or in any private street, the consent of the occupier or manager must be obtained and compensation must be paid.

9.3.2.2 The Gas Act 1986
This Act applies to all public gas transporters. Such companies have the power to break up any street, as defined in the NRSWA 1991, to instal pipes and undertake any associated works (Schedule 2), but where the street is not dedicated to public use, the right is restricted to the giving of a supply to a property which abuts that street. Where a property has been sold away from a farm with the right to use a track for access, this power allows the laying of a gas supply under that track.

9.3.2.3 The Electricity Act 1989
Under this measure, all licensed companies (National Grid Company plc and the regional electricity companies in England and Wales) have the power to instal apparatus in any street which is maintainable at the public's expense.

9.3.2.4 The Water Industry Act 1991 and the Water Resources Act 1991
These both give all water and sewerage undertakers a wider power to install pipes in or under any street, and the NRSWA 1991 definition is unqualified.

9.3.2.5 The Pipelines Act 1962
This Act permits a pipeline to be laid in a highway which is maintainable at public expense with the consent of the highway authority, in a street which is to become maintainable at public expense with the consent of the highway authority and the manager of the street, and in any other street with the consent of the manager.

9.3.3 Works in the highway
The right to undertake works in the highway is now prescribed in the New Roads and Street Works Act (NRSWA) 1991. This requires operators intending to do works to apply to the highway authority for a licence to do works. Once granted, the operator must

- Give at least 7 days' notice to the authority and all other operators whose equipment may be affected
- Comply with any requirement of the highway authority to reduce disruption anticipated as a result of the works

- Take appropriate safety measures and take special account of those with disabilities
- Avoid unnecessary delay
- Reinstate the surface to designated standards

These rules can have a consequential effect: in most cases where a highway benefits a property or business, the owner must accept both the benefit of that access and the disturbance which may arise when work is undertaken in the highway, whether to resurface it or to dig it up and instal services. The principal exception covers the water and sewerage companies who do have a duty to pay compensation; in addition, the Gas Act was amended in 1995 to provide for compensation in some cases. Both of these are explained in the relevant sections below.

There may be a remedy under the McCarthy Rules (see section 8.4.13), but that is unusual. The rules are different under the Gas Act 1986, and the matter is being considered by government within the continuing review of CPO legislation.

9.4 Electricity

9.4.1 Electricity supplies

In England and Wales, the National Grid Company plc (NGC) has the responsibility for conveying electricity at high voltages from power stations into the network of the regional electricity companies (RECs), which then deliver the power to consumers. NGC has the obligation to achieve a balance in the system, by anticipating overall demand and arranging the supply from power stations to meet the needs of all RECs. This is achieved by generating power in different ways, because once electricity has been generated it must be delivered into the system. Some power sources can respond at short notice and are used to meet peak demands and others cannot respond quickly and these are primarily used to meet the core electricity demand.

RECs secure their primary supplies from NGC. They also take supplies directly from renewable schemes, and are obliged to do so from schemes approved under the Non-Fossil Fuel Orders. These supplies are less predictable as wind turbines, hydro schemes and others all rely on natural factors to generate power, but it is for each REC to allow for that in balancing supply and demand in its area.

9.4.2 Acquiring rights in land

When considering issues relating to land, the electricity industry in England and Wales can be divided into three sections: the generators, the *National Grid Company plc* (NGC) and the *regional electricity companies*

(RECs). Other companies act as suppliers of electricity but they rely on the apparatus of NGC and the RECs to distribute the power.

The Electricity Act (EA) 1989 includes the right for operators to acquire land by seeking a CPO (Schedule 3). This normally applies when the use of a specific area of land is required, for example for a power station. The normal rules for compulsory purchase apply with minor amendments, and compensation is assessed on the normal basis (see Chapter 8).

CPOs can also apply to the acquisition of land for substations, but many of these are situated on land held on long leases. When a lease expires, the company will normally wish to renew; it is probable that such leases have the protection of the Landlord and Tenant Act 1954, with the right for the tenant to seek a new lease; equally under the Act the power for the landlord to object to a new lease can arise, particularly where he can show that he wishes to occupy the site himself or redevelop the premises. For this reason, the 1954 Act rarely features in discussions on renewal, and the parties proceed on the basis that compulsory powers could be pursued if there is no agreement.

The EA 1989 also gives power for companies to acquire other rights (Schedule 4). Most importantly, this includes the power to acquire 'the necessary wayleave' in order to instal electricity lines on poles or pylons over the land, or, less often, underground.

The companies must seek to negotiate a wayleave with the owner and occupier along the most acceptable route. This will include the offer of compensation for installing and keeping the apparatus on the land, and for the disruption to farming operations. While the majority of all rights in land are secured by agreement, it must be remembered that owners' and occupiers' rights are limited by the rules laid down in the 1989 Act.

Wayleaves created by agreement usually run for an initial term, of perhaps five years, and then run on from year to year.

Where a company fails to secure the rights by agreement, it must serve notice on the landowner and the occupier along with a consent form; the notice must give 21 days notice for the return of the consent form. If the form is not returned, the company may, within three months, make an application to the Secretary of State at the Department of Trade and Industry for 'the necessary wayleave'. If the owner or occupier objects formally to that application, a hearing must be held. If the inspector accepts that the apparatus should be installed along the proposed route, a necessary wayleave will be granted for a period of years, and it is almost always the case that the period will be 15 years. The rules for such hearings are set out in the Electricity Generating Stations and Overhead Lines (Inquiries Procedures) Rules 1990 (SI 1990/528) as amended.

The Secretary of State cannot grant an order for the necessary wayleave where the line crosses land covered by a 'dwelling', being a house and its private garden, or land which has been granted permission for housing which is likely to be developed. The effect of an order is to permit the company to instal and keep the apparatus on that land for the 15 year period. At the end of this time, the owners and occupiers have the same right to terminate the wayleave. Once the apparatus is installed, the electricity companies are able to undertake works to protect and maintain it.

9.4.3 *Underground cables*
For high voltage supplies, a permanent easement will normally be acquired but for lower voltage cables, wayleaves can also apply to underground routes. The principles are exactly the same, but the underground supplies do represent a hidden hazard. Owners and occupiers are liable if an underground supply is disturbed or broken when working on the land.

9.4.4 *Terminating a wayleave*
The owner and occupier of land have the right to terminate the wayleave by giving notice under the agreement, usually 6 or 12 months, but most have a longer initial period. When that notice has expired, a further notice requiring the removal of the apparatus must be served on the electricity company (paragraph 8 Schedule 4). At that time, the company can serve a counternotice seeking to retain the apparatus or make a CPO; if a CPO is made the usual procedures apply and it can only be implemented if it is confirmed (see also Chapter 8). If agreement cannot be reached, the matter will be taken to a hearing and either the apparatus will be moved or the necessary wayleave will be granted.

Where land changes hands, the new owner is not bound by the terms of the existing wayleave. He can receive the payments for the presence of and disturbance caused by the apparatus but the wayleave agreement entered into by the previous owner is not binding. Accordingly, the new owner does not need to serve a notice to terminate the wayleave; he can simply move to the second stage outlined above, and serve a notice to remove the apparatus.

Most commonly, a wayleave will be terminated when an owner wants the line moved to permit development or to improve a view. The compensation payable if the line is kept can be substantial and this can result in an agreement for a new route, and the movement of the line. Where the line is retained, the compensation reflects the full loss to the owner: where the loss arises because the presence of the lines has restricted the amount of development permitted in a planning consent, serious problems can arise when seeking to establish the exact nature of the planning permissions that would have been granted if the lines had not existed. This is because planning permission will normally only be granted where the development is likely to be implemented.

9.4.5 *Compensation*
The payment for the right is divided as follows.

- A payment is made to reflect the injurious affection caused to the land arising from the presence of the apparatus. This is usually called 'rent' for the land, although it is strictly compensation.
- A second payment is made to cover both the additional costs and losses caused as a result of having the apparatus on the land; for farmers, this is calculated on a national basis for both arable and grassland farms, using the *ADAS formula*. This flows from a major study of the costs and losses undertaken by ADAS in 1994/5 and updated each year with the benefit of the annual census and other statistics. In special cases, e.g. high value fruit growing or double-cropped land, the payments can be adjusted.
- Compensation is also paid for any losses incurred when the apparatus is installed or when any damage is caused when other work is undertaken.

Compensation for a wayleave can be paid as a lump sum or by periodic payments, or a combination of the two (paragraph 3 Schedule 4). In this way, a payment can be claimed for the right to retain the apparatus on the land but an annual payment can still be taken for the disturbance to operations on the land. Where a necessary wayleave has been granted, the owner can be restricted in his ability to develop or use his land for a particular purpose during the fixed term; accordingly, the landowner or occupier may be entitled to claim substantial compensation for the grant of a right for 15 years, similar to the payment of a premium at the start of a lease.

Where a CPO is obtained and exercised, or a permanent easement has been granted for the apparatus to remain permanently on the land, the compensation is paid as a single payment. The normal rules apply to the assessment of compensation (see Chapter 8).

9.4.6 *Telecommunications use of electricity apparatus*

The electricity companies can also operate fibre-optic networks for telecommunication purposes. Where the service is solely concerned with the operation of the electricity business, the rules set out in the Electricity Act 1989 apply; where there is any third party use, the Telecommunications Act 1984 applies. People can be concerned that they will not know whether a third party is using the wires, but it is an offence to operate such a business without a licence, and the regulatory bodies are in place to monitor and investigate such matters.

9.4.7 *Power to survey*

Companies have the power to enter land for the purpose of survey and establishing whether the subsoil is suitable (paragraph 10 Schedule 4). To do so, 14 days notice must be given. It is an offence to obstruct any person who can show proper authority to be on the land after the giving of proper notice.

9.4.8 Lopping of trees

Companies have the power to require the lopping of trees which interfere with, or cause a hazard, in the vicinity of the overhead lines (paragraph 9 Schedule 4). When this is proposed, 21 days notice should be given requesting the owner or the occupier to undertake the work; the company should pay the reasonable cost of carrying out the work. If the work has not been done within 21 days and the owner or occupier has not objected to the work, the company may enter and undertake the work itself. If there is an objection, a hearing must be held to decide whether the work set out in the notice should proceed.

9.5 Telecoms

9.5.1 Telecommunications Code

The rules which allow a telecommunications or cable TV company to acquire rights in land are set out in the Telecommunications Code (set out in full in Schedule 2 to the Telecommunications Act 1984). To use these rights, the operator must be licensed as a *Code systems operator.*

The arrangements here are closer to the open market than exist in any other utility industry, and it is rare for the compulsory powers to be used.

9.5.2 Acquiring rights in land

Operators have the power to seek a CPO to acquire land or rights in land but this is subject to both the consent of the Secretary of State (DTI) and the approval of OFTEL (section 34 TA 1984). It appears that no such CPO has been sought since the industry was privatised. This is particularly the case on telecom towers, where a market operates without compulsion, both on a leasehold and freehold basis.

Usually a wayleave is agreed for the installation of overhead or underground lines. This agreement has to be given by the occupier, even where the terms of the lease reserve the power to grant such rights to the owner. Where the occupier gives consent, the owner (or head lessee or other superior interest) is not bound and it is usual for his consent to be obtained also. When the consent of an owner or occupier is given, it is binding on the successors to that interest, but not to a superior interest (paragraph 2 of the Code).

Where the installation of apparatus affects a third party, e.g. by restricting a right of access, that person must also give his consent (paragraph 3).

Where an occupier or owner refuses to give consent and the operator cannot provide the service at reasonable cost without such consent, the operator can seek an order to instal the apparatus, similar to a wayleave. The operator must submit a consent form and request its return within 28 days; if the form is not received, the operator can apply to the county court for an order. The court must

(1) consider whether the grantor will be adequately compensated for his loss, taking account of the benefit to the person receiving the service;
(2) take into account the government view that no-one should be unreasonably denied access to a telecommunications system; and
(3) decide the terms, including payment, on which the order can be exercised, see section 9.5.4 below, (paragraphs 5–7 of the Code).

Where a person has requested a telecom supply, he can require the company to seek an order if the owner of any land refuses to grant the rights (paragraph 8).

Operators also have the power to fly lines over land where there is no need to erect a pole or other structure on that land (paragraph 10). The lines must be at least 3 m above the ground, 2 m from any building and the lines must not interfere with any business undertaken on the land.

Any person has the right to object to the installation of new apparatus within three months of its erection where the apparatus will interfere with a proposed improvement (paragraph 20). The apparatus will be removed if the interference can be proved and if the removal does not substantially interfere with the operator's service.

9.5.3 *Terminating an agreement*

An owner or occupier can seek to alter the route of a supply when it interferes with a new use or the development of his land; unlike electricity, this applies whether a permanent right or a wayleave exists (paragraph 20). In other circumstances, and where this is not the case and the wayleave agreement provides for the right to be terminated, the owner or occupier can serve notice (paragraph 21). In either of these circumstances, the operator can apply to the court for an order to keep the apparatus in place. If successful, the operator will acquire a permanent right in the land but full compensation will be due; compensation is assessed on the standard basis and settled by the Lands Tribunal in the event of a dispute.

Where apparatus is abandoned, the operator has a duty to remove it. In the case of underground supplies, operators can come to a separate agreement with the owner to leave the ducts and wires in the land and making a payment. The operator has a duty to remove the apparatus, reinstating the land and paying compensation for the damage caused, if agreement cannot be reached.

9.5.4 *Payment*

With telecommunications, as distinct from all other utilities, there is a different basis for payment. The Telecommunications Code requires the court to fix the terms on which any right can be exercised. This includes the 'consideration', which must be fair and reasonable and based on the assumption that the arrangements have been entered into willingly.

This has exercised the courts on the few times that it has been considered. In the *Mercury* case (*Mercury Communications Ltd* v. *London &*

India Dock Investments [1995] 69 P&CR 135), the judge expressed concern about the responsibility being placed on the court, when there was no benchmark to assess the payment. He then calculated the consideration by adjusting the payments made under another agreement which had been settled by the same parties.

9.5.5 Cable companies

The same rules apply to cable companies provided they are Code systems operators, but attention is drawn to the requirement to obtain consent and pay compensation for cables laid in private streets and tracks.

9.6 Gas

9.6.1 The gas industry

The industry is now divided into three parts: the supply of bulk gas (for example from gas fields and former coal mines) by *gas shippers*, the transport of gas through the pipeline network by *gas transporters*, and the actual supply of gas to consumers by *gas suppliers*. Gas suppliers are the main market makers as they acquire the gas from gas shippers and rely on gas transporters to deliver it to the end users, their customers.

BG Transco is the only public gas transporter currently licensed to operate a pipeline network under the Gas Act 1986. In order to allow shippers and suppliers to operate, BG Transco can be required to accept gas of a similar kind into the system and to deliver it elsewhere. The gas must comply with prescribed standards as to its properties, condition and composition; the gas transporters must also operate to standards, covering the purity of gas within the pipelines and pressure. A regulator supervises the industry to ensure that the rules are obeyed, that BG Transco does not discriminate between companies, and that the price for carrying the gas is reasonable.

BG Transco has duties to

(1) develop and maintain an efficient, coordinated and economic system for gas supply;
(2) transport gas within this system on behalf of gas shippers, involving the receipt, transmission and supply to end users' service pipes of gas supplied to it by other licensed companies;
(3) maintain, repair or renew the pipeline network, including service pipes; and
(4) connect any premises (where the likely gas consumption does not exceed 75,000 therms in any 12 month period) where the premises are in an authorised area, and within 23 m of a main or which could be connected to the main by a service pipe laid by the owner. (The duty to connect is subject to safeguards to ensure that there is a

minimum level of supply and that the terms are reasonable, including payment for the connection.)

Not all the pipelines built by BG Transco are covered by the Gas Act. Some are built to supply a single project, e.g. a new power station, and are not part of the national gas transmission network. Such schemes are covered by the Pipelines Act 1962, (see section 9.9).

9.6.2 Acquiring rights in land

A public gas transporter has the power to acquire land and rights in land by CPO (section 9 and Schedule 3 GA 1986). The normal rules apply for the making of a CPO and for hearing all objections.

At the same time, the transporter must obtain a *pipeline construction authorisation* (PCA) from the Secretary of State at the Department of Trade and Industry. The PCA will normally be accompanied by the grant of planning permission; for lesser pipelines and other limited works, gas transporters have deemed planning permission under Part 17 of the General Permitted Development Order, subject to giving at least eight weeks notice to the LPA.

An application for a PCA must be accompanied by a book of reference with the names of the people who will be affected along the route and details of all their interests. It must also state the number of consents which have been granted by agreement, and details of those which have yet to be acquired; the applicant must also show that it has taken reasonable steps to acquire those outstanding interests.

Most rights are acquired by agreement, and the majority are in accordance with an agreement between BG Transco and the CLA and NFU in England and Wales, and the Scottish Landowners' Federation and the NFU in Scotland.

9.6.3 The basis of compensation

The rules set out in the Acquisition of Land Act 1981 and the Compulsory Purchase Act 1965 apply, with minor amendments to allow for the creation of new rights. Compensation is assessed on the standard basis, with the payment based on the loss to the grantor, in accordance with section 5 of the Land Compensation Act 1961, compensation for injurious affection and disturbance is also payable (see Chapter 8).

9.6.4 The Minerals Code, development and planning

As with any other CPO, the minerals in the land are only acquired by the gas transporter if they are specifically included in the CPO. Normally, the CPO is silent and the former owner retains the minerals and the right to work them. The Minerals Code is set out in Schedule 2 to the Acquisition of Land Act 1981.

If the owner secures planning permission to work the minerals, he must give the gas transporter 30 days notice of the intention to withdraw support from the pipeline. The gas transporter can serve a counternotice requiring the owner to leave the minerals in the ground; compensation has to be paid for the value of the sterilised minerals along with compensation for all consequential losses, e.g. the increased cost of operating the plant and equipment without the throughput of these minerals. The compensation is assessed at the date when the minerals would have been worked, not at the date of the original acquisition.

The standard agreement with BG Transco also includes a development clause to the same effect. In the event that planning permission is granted for a use of the land which is covered by the agreement, the owner can give notice requiring the pipeline to be moved so that it follows another route on his land, but still linking up with the same points on the boundary so as not to require a move also on the land belonging to a third party. It is then for BG Transco to decide whether to reinforce the pipe so that the development can proceed, or move the pipe, or to retain the pipe as it is and pay compensation for the loss.

The difficulty with both minerals and development is the securing of planning permission on the route of a high pressure gas pipeline. The majority of new developments are prescribed in local development or local mineral plans. If land is not designated for development in these plans, applications are generally refused. LPAs are also concerned to allocate land for development which can actually be developed. For that reason, land which is crossed by a high pressure pipeline is not usually allocated for the working of minerals or for the development of housing or other buildings; accordingly, it can be very difficult to invoke the Minerals Code or development clause. This is a 'Catch-22' problem, which has yet to be resolved.

There is a further complication caused by the use of a pipeline for the carrying of a volatile substance under pressure: the Health and Safety Executive (HSE) is responsible for ensuring that people are protected from risk and takes a close interest in proposals which are close to pipelines. That interest can extend to land outside the area covered by the agreement, and it can extend several hundred metres from the line of the pipe. This can mean that development proposals, e.g. to convert a barn to a house or to build a business park, are refused on the grounds that the presence of the pipeline is a danger and that in the event of damage, the occupiers would be at risk of death or injury.

The date for assessing compensation on a CPO is the date of agreement, the date of entry or the last day of the Lands Tribunal hearing. This means that proper account is only taken of the planning permissions which exist at that date, along with any realistic hope value that consent will be granted in the future. If circumstances change after the pipeline is constructed, the owner of the land may be prevented from undertaking a development with no redress against the gas transporter.

This issue can also affect people who own land close to the line of a pipeline, even though their land has not been touched by the construction works. Such people have exactly the same right under the Minerals Code but they have no rights in relation to any development scheme which is frustrated.

9.6.5 Compensation for works in streets

Some utility companies have a right to lay apparatus in streets without liability for any loss caused. The position for gas changed in the Gas Act 1996 when a new rule was introduced to provide for compensation to small businesses. The Gas (Street Works)(Compensation for Small Businesses) Regulations 1996 require compensation to be paid where

- the street works last more than 28 days and cause loss to a business;
- a claim is submitted within three months of completion of the works with full details being provided within six months of completion;
- the turnover of the business, along with any associated businesses, does not exceed the threshold, which is set at £1 million plus indexation in line with the RPI from 1 March 1996; and
- the loss is at least £500 and more than 2.5% of turnover.

9.7 Water resources and drainage

9.7.1 The Environment Agency

In England and Wales, the *Environment Agency* (EA) is responsible for the management of water resources. This covers the main river system but does not extend to the drainage of all land, where the duty vests in internal drainage boards or the local authority. The Water Resources Act (WRA) 1991 applies and all statutory references in this section are to that Act unless otherwise stated.

The duties of the EA in managing water resources include

(1) conserving, redistributing or allocating water resources, including the ability to grant abstraction licences (subject to exemption for small amounts as set out in section 27 WRA), the request to the Secretary of State to make a drought order to restrict usage, and the determining of minimum river flows;
(2) making water resources management schemes with water companies;
(3) making arrangements for flood defence; and
(4) controlling pollution of water resources.

In addition, the EA has powers to undertake further works for flood defence and drainage purposes on main rivers, including

(1) maintaining and improving existing works, and the construction of new works, for the purposes of land drainage; and
(2) similar works for the purposes of sea defence or containing tidal waters.

9.7.2 Acquiring rights in land

The Environment Agency has the power to acquire land or rights in land by CPO (section 154). The Acquisition of Land Act 1981 applies to the acquisition. Where rights only are acquired, the Land Compensation Act 1965 is modified to ensure that the compensation is in line with the normal principles (see Chapter 8).

As an alternative in some circumstances, the EA can seek a *compulsory works order* (CWO) where the work involves building or engineering works or the discharge of water into any inland waters or underground strata (section 168).

More usually, the EA will exercise its right to enter land and lay pipes: section 159 applies to all streets, including private roads and tracks, and section 160 applies to private land. This power is granted for the laying of either

(1) a *resource main*, which is defined, in effect, as the bulk supply of water, but not for its delivery to premises which is covered by the Water Industry Act 1991, see section 9.8; or
(2) a *discharge pipe*, which means any pipe from which water is discharged when the EA is constructing, altering, repairing, cleaning or examining any reservoir, well, borehole or other work used in the exercise of its functions.

When laid, and to avoid any legal uncertainty, such pipes are specifically vested in the EA (section 175).

Further, works associated with such pipes include

(1) the provision of any drain or sewer to intercept, treat and dispose of any foul water on any land, or otherwise preventing the pollution of any water;
(2) the construction of any watercourse for any of the same purposes; and
(3) not only laying and maintaining such a pipe, but also cleaning it, altering its size or its route, moving it, removing it, or replacing it with a different sized pipe if that is to be used for the same qualifying purpose.

The power to lay qualifying pipes is subject only to the requirement to give reasonable notice, except in an emergency. In the case of new works this is required to be at least 3 months; for alterations to existing pipes, the period is 42 days. Seven days notice of entry must also be given.

There is no procedure by which anyone can object to such pipe-laying works but if access to the land is denied or the EA believes that access will be obstructed, it can apply to the magistrates' court for a warrant. A warrant can only be issued if the magistrate is satisfied that access has been, or will be, refused, and if the EA has given notice of the application for a warrant to the occupier. The magistrate must also be satisfied that the EA has the proper authority to undertake the proposed works. The court hearing is the only opportunity to object to the works: it is questionable whether these powers are compatible with the Human Rights Act 1998, but the matter has yet to be tested.

The EA has the power to use any watercourse to take the water from a discharge pipe, provided that it does not affect any railway, highway, or property of a navigation authority. Importantly, this power is restricted so that the EA cannot use a pipe of over 229 mm (9"), except in an emergency, without the consent of all persons within three miles downstream of the discharge point; if consent is not forthcoming, the matter must be decided by an arbitrator appointed by the Institute of Civil Engineers.

9.7.3 Compensation

Compensation is payable in most circumstances. The normal rules apply on the exercise of any CPO (see Chapter 8). Where the CPO involves the taking of newly created rights in land, the normal rules are modified to ensure that the statutory principles for assessing the compensation are maintained (Schedule 18).

Compensation is also payable for works undertaken under any compulsory works order (schedule 19): in the case of building and engineering works, it is on an equivalent basis to that which would have applied if the land had been acquired by CPO. Any person who suffers loss as a result of the discharge of water through a discharge pipe is entitled to compensation. Those who suffer loss as a result of the abstraction of water or obstruction to the flow are not entitled to compensation (paragraph 7 Schedule 19).

Schedule 21 sets out the arrangements for assessing compensation when pipe-laying powers are used to instal any pipe or watercourse in any street or private land. When pipes are laid in private land, compensation is paid to cover the depreciation in the value of the land, together with any other loss or damage suffered by the owner or occupier (paragraph 2). Paragraph 4 provides for compensation for losses caused by the discharge of water (there is a specific obligation for the EA to do as little damage as possible), and attract compensation under losses caused by flood defence and drainage works (paragraph 5).

Disputes are settled by the Lands Tribunal, except where there is a dispute following the discharge of water (section 163) when the matter is settled by an arbitrator appointed by the Institute of Civil Engineers.

9.7.4 Land drainage

The Land Drainage Act 1991 can also affect landowners. There are provisions which permit

(1) the drainage board or local authority to require a person to clean out a watercourse to facilitate drainage (section 25);

(2) the owner of land to seek authority to undertake work on neighbouring land to allow his land to be drained (section 22];
(3) an owner to secure an order from the Agricultural Lands Tribunal requiring a neighbour to clean out his ditch (section 28); and
(4) drainage boards and local authorities to acquire land or rights in land for the proper exercise of their duties (section 62).

In each case where a third party is affected, there are requirements for the service of notice and the hearing of objections. The basis of compensation for the acquisition of land or rights, is on the usual basis (see Chapter 8).

9.8 Water and sewerage

9.8.1 Powers of entry

Along with the Environment Agency in the discharge of its duties in relation to water resources and management as set out above, the licensed water and sewerage undertakers providing these services have the greatest powers to enter land and do works; the arrangements have been carried forward from the Public Health Acts of the nineteenth and twentieth centuries when there was a real need to improve the supply of potable water and efficient and safe drainage. In this section all statutory references are to the Water Industry Act 1991 unless otherwise stated.

The activities are divided and in some areas, the water supply and sewerage functions are undertaken by different companies. With the introduction of competition, some customers can secure water supplies from companies outside their immediate area, under arrangements which allow shared use of the pipes.

9.8.2 Water companies

Water undertakers have statutory duties to

(1) develop and maintain an efficient and economical system for the distribution of water within its area;
(2) provide water to premises and people when required; and
(3) maintain pressure at a level which ensures that all buildings can be served by the system.

Importantly, for customers and landowners, any person who owns or occupies property and some public authorities can requisition a new water supply (section 41). The undertaker must comply within three months of the date when the terms for providing the connection are agreed, unless the period is extended by agreement or at arbitration. The undertaker can impose reasonable financial conditions taking account of the cost and the projected usage. If there is a delay and the person requisitioning the supply suffers a loss, he can claim compensation but the undertaker can defend itself by showing that it has taken all reasonable steps to comply with the demand.

A householder can require a supply to be laid from the nearest main to his property for domestic purposes and the undertaker can recover the reasonable cost. A non-residential supply can also be requested but the undertaker does not have to make a connection if it would incur unreasonable expenditure.

9.8.3 Sewerage companies

Sewerage undertakers have a duty to

(1) provide, improve and extend the system of public sewers and to keep them clean in order to drain the area; and
(2) empty the sewers and do such other things as are necessary to dispose of the contents of the sewers.

A sewer is defined so as to include all drains and sewers which drain buildings and their yards. A sewer does not have to carry foul sewage but can carry storm or flood water. Not every drain is a sewer, as there are many private drains and private sewerage pipes which are the responsibility of the users who benefit from them.

The sewerage undertaker is responsible for the pipes but not for what flows through them, because the sewage comes from third parties. If a sewer blocks and overflows, there is no claim for the consequential loss unless it can be shown that it is the design rather than the contents which have caused the overflow. This is distinct from water pipes where the water undertaker is liable as it supplies and controls the water in the pipe, yet the basis of assessing compensation for the loss to the owner of the land is the same for both water pipes and sewers.

Any person can require the undertaker to connect a private drain or sewer to the mains system, unless the condition is such that the connection would prejudice the general operation of the system. A person can also requisition a sewerage undertaker to provide a sewer for the drainage of property for 'domestic purposes' (section 98), where this means the carrying away from a building and land occupied with the building of the contents of any lavatory, water used for cooking or washing, or surface water (section 117).

The work must be completed within six months of the date on which the person requisitioning the service has agreed to pay the reasonable cost of the connection, unless this period is extended by agreement or at arbitration. That cost is calculated on the basis of the cost of the works and the income to be derived from the service towards that cost, and the 'relevant deficit' is payable, as a lump sum or in 12 annual instalments.

Local authorities also have a role to play: where a property is already connected to the sewer, no alterations can be made without giving 24 hours notice to the local authority. Where an existing system is inadequate, defective, causing a nuisance, admitting soil water or is otherwise in a condition which is likely to be harmful to health, the local authority can require the necessary works to be carried out to put it in a reasonable condition.

9.8.4 Acquiring rights in land

Both water and sewerage undertakers have extensive rights to acquire land, although they are not the same in all respects.

The undertakers have the power to make and pursue CPOs, and these can include the acquisition of new rights created in the land. The Acquisition of Land Act 1981 applies to the acquisition. Again, where rights only are acquired, the Land Compensation Act 1965 is modified to ensure that the compensation is in line with the normal principles (see Chapter 8). The undertakers can also seek compulsory works orders, equivalent to those under the Water Resources Act, and Schedule 11 sets out the procedures that apply.

More usually, the companies use the power to enter land and lay pipes (section 159) in order to

(1) Lay and keep a relevant pipe in any land other than a street (in section 158)
(2) Inspect, maintain, adjust, repair or alter any such relevant pipe
(3) Carry out works requisite for, or incidental to the purposes of any of the above works

One undertaker has used the power in (3) above to take and use land as a contractor's depot while constructing a pipeline on other land nearby, even though the pipe was not laid in land owned or used by the farmer. It is doubtful that this is permitted under the 1991 Act, and almost certainly it is not under the Human Rights Act 1998: such occupation interferes with the peaceful enjoyment of the property, and it is the actual laying of the pipes which has merited the special regime of 'serve notice and enter'. As for the Water Resources Act, if the undertaker has been denied access after due notice, or believes that this will be the case, the undertaker may apply to the magistrates' court for a warrant (see section 9.7.1 above).

A relevant pipe is defined as

(1) for a water undertaker as a water main (including a trunk main), resource main, discharge pipe or service pipe; and
(2) for a sewerage undertaker as any sewer or disposal main.

The undertaker must give proper notice and this is defined as three months in the case of new works, 42 days for alterations to existing apparatus, and 'reasonable notice' in the case of emergency or where a supply has been requisitioned (sections 41 and 98).

An important difference between water and sewerage undertakers was highlighted by the Court of Appeal in *British Waterways Board* v. *Severn Trent Water Ltd* (judgment 2 March 2001). Water undertakers have the power to discharge water into any water course (section 165), but sewerage undertakers do not. The court held that it would be wrong to imply such a power merely because sewerage undertakers have pipe-laying powers under section 159; they will have to route their pipes to locations where they have or can acquire the necessary right to discharge. This case illustrates the need to understand the proper basis on which any powers are exercised, and to review the actual wording of the statutes when involved in a dispute.

9.8.5 Compensation

On the exercise of a CPO the normal rules apply. When the undertaker has served notice and used the powers in section 159, the compensation is assessed in accordance with schedule 12. Compensation is paid for depreciation in the value of the land, as well as for any other loss or damage suffered as a result of the works; injurious affection is also paid for the loss of value of any adjoining property but full account is taken of any increase in value arising out of the works. In *Collins* v. *Thames Water Utilities Ltd* [1994] 49 EG 116 before the Lands Tribunal, compensation was settled for the laying of a sewer at the bottom of a garden. Compensation was paid for entry, disturbance and the time spent by the claimant, but nothing was paid for the effect of the continuing right to enter to maintain repair and replace the sewer, as the property had been connected to, and benefited from, the sewer.

As a general rule, compensation is paid for half the value of the land affected by the pipe; it is usual for the undertaker to specify the width which will be permanently affected by the presence of the pipe, but this is usually substantially narrower than the strip occupied to lay it. There is no statutory formula for this and it is a matter for discussion. Disputes are settled by the Lands Tribunal.

The lead case in settling compensation is *St John's College Oxford* v. *Thames Water Authority* [1990] 1 EGLR 229. The Lands Tribunal had to interpret the Water Act 1945 but the principle remains the same. The compensation must be assessed on a 'before and after' basis, and take account of the fact that the land through which the pipe is laid is affected and cannot subsequently be built on. The property must be assessed as it stands, and the value of the land in this case was reduced because it was subject to a secure Agricultural Holdings Act 1986 tenancy. This case also held that no legal right was created in the land, but this is no longer at issue as any works carried out in the exercise of statutory powers are vested in the undertaker [section 179].

There are several other matters of note on compensation.

(1) Interest is payable: this is not in the Act but was decided in *Taylor & another* v. *North West Water Ltd* [1995] 1 EGLR 266. The court held that the pipe-laying powers amounted to a compulsory acquisition and that interest was properly payable in order to ensure that

the claimant received full compensation, but interest only accrues from the time the claim is made.
(2) The undertaker is responsible for the actions of its contractor when pipe-laying works are carried out, on the grounds that the contractor is acting as the agent of the undertaker which has the power to enter under notice. This was established by the Lands Tribunal in *Donovan v. Dwr Cymru* [1994] 1 EGLR 203, where a sewer was laid on land outside that covered by the notice but it was held that the land was 'relevant land' and the pipe was a 'relevant pipe'.
(3) Unlike other utilities, save for gas as explained in section 9.6.4 above, water and sewerage undertakers are obliged to pay compensation to any person who has sustained loss or damage (section 180 and Schedule 12). This includes those who suffer loss of business as a result of works in the public highway, even though none of their own land is affected.

9.8.6 Development

Where land is crossed by a water pipe or sewer, the owner can require the undertaker to alter or remove the pipe or sewer. The undertaker must comply unless the request is unreasonable, for example because an alternative design would accommodate the same development yet allow the pipe or sewer to remain. The landowner is required to meet the reasonable cost of altering or removing the pipe.

When land subject to a CPO has an enhanced value because it controls the access to other developable land, the enhanced value is taken into account in the compensation. It has been suggested that this should apply also where water mains or sewers have to be laid across private land. This is, however, not the case as any developer of the land can requisition a supply and the water and sewerage undertakers are under a duty to comply; this removes the need for any developer to enter into private negotiations, unless the works fall outside the scope of the statutes.

9.9 All other pipelines

The Pipelines Act 1962 applies to cross-country pipelines for the conveyance of most materials, except for air or water in any form. Other Acts also apply to pipelines including, for example, the Requisitioned Land and War Works Act 1948 which covers government oil pipelines.

Under the 1962 Act, a pipeline construction authorisation (PCA) must be obtained from the Secretary of State at the DTI for any pipeline over 16.09 km (10 miles) and the same principles apply, as for the Gas Act (see section 9.6.1 above). When the application is made, the company must give notice to the LPA and to any other persons, as the Secretary of State may specify. If the LPA objects, a public inquiry must be held; if any other person objects, an inquiry can be held or other arrangements can be made for the objection to be heard and considered. The Secretary of State can also assess whether the construction is necessary or appropriate given the duty on all applicants to lay pipes in such a way that they reduce the need to lay others (section 9 PA 1962). When a shorter pipeline is to be constructed, notice must be given to the Secretary of State. When a short pipeline is to be joined to another short pipeline so their combined length exceeds 16.09 kilometres, then application must be made to the Secretary of State. Once a PCA has been granted, work must have been started to a substantial extent within 12 months or the PCA will lapse, unless the Secretary of State grants an extension of time.

A pipeline company can ask the Secretary of State for the power to secure land, by compulsory purchase (section 11 PA 1962), or rights in land, with a compulsory rights order (section 12). If the application is allowed to proceed, the company must serve notice on all owners and occupiers; anyone who lodges an objection must be given an opportunity to present their case, at inquiry or otherwise, and then it will be decided whether or not to confer CPO powers. The procedure for determining CPOs and CROs is laid down in Schedule 2 PA 1962.

A CRO can include not only the right to instal a pipeline (including maintaining, repairing and renewing it), but also ancillary rights

(1) to put and keep markers on the land;
(2) to erect stiles, gates, bridges, and culverts to facilitate access; and
(3) a right to place other materials, plant or apparatus on the land; but
(4) the land subject to (1) and (2) above must be specified in the order.

Otherwise and subject to minor amendment, the normal rules for compulsory purchase apply on the exercise of a CPO or a CRO (see Chapter 8).

Chapter 10
Liability of the Occupier of Land

Abbreviations in this chapter
'1957 Act'   Occupiers' Liability Act 1957
'1971 Act'   Animals Act 1971
'1977 Act'   Unfair Contract Terms Act 1977
'1984 Act'   Occupiers' Liability Act 1984
'Secretary of State' Secretary of State for the Environment, Food and Rural Affairs, or the National Assembly for Wales where the context requires

10.1   Impact of land use

The use of land can have a far reaching impact – on invited visitors, neighbours, the endless band of officials authorised to enter land, trespassers and the general public. A glance through the law reports is enough to demonstrate the almost limitless ways in which unpleasant happenings can lead to claims against the occupier of land. The courts and the legislature try, with varying success, to balance the scales of justice between people who suffer from injury, loss or discomfort on or near the land of others and the occupiers who can reasonably expect them to take some care of themselves.

This chapter summarises the legal duties owed to others by occupiers of land. It is mostly concerned with the laws of negligence, nuisance and animal trespass.

10.2   Insurance

The importance of taking out adequate insurance against civil liability cannot be overstated. Accidents will happen, and they can be very expensive. It is also vital to inform the insurers of any change in circumstances which may affect the risk insured against, otherwise the insurers may be able to avoid the policy in the event of a claim. Under the Employers' Liability (Compulsory Insurance) Act 1969 (as amended) an employer is required to take out insurance for claims for compensation for accidents at work pursed by employees in civil actions. The current prescribed cover is £5 million. Although it is usual for most policies to offer cover of £10 million, and sometimes more. Certificates of insurance are required to be retained for 40 years. In an increasingly litigious world, it is also advisable to include cover for legal expenses.

10.3 Negligence

Negligence is the breach of a duty of care, owed by one person to another, which results in damage to that party. Therefore a person will not be liable for any act or omission which results in damage to another – it has to be shown both that there was sufficient proximity between the persons (e.g. both are road users) for a duty to be owed, and that that duty has been breached. The standard of duty is that of an ordinary reasonable man – this is an objective test, and whilst it does not expect every person to be perfect, it does not take account of idiosyncrasies.

It is essential that the complainant has suffered damage for there to be an actionable case. Damage here generally means physical harm to persons or property, but it may include consequential economic loss caused to someone, and may be caused by act or omission.

10.4 Occupiers' liability

10.4.1 The legislation

The Occupiers' Liability Act 1957 sets out the legal duty owed by the occupier of premises to *visitors* (as defined in the Act) and the Occupiers' Liability Act 1984 the duty owed 'to persons other than his visitors'. 'Premises' in the Act does not have its colloquial meaning. It includes land and waters and in effect means the property of the occupier. Other specific legislation covers employees and landlords.

10.4.2 Visitors and non-visitors

'Visitors' under the 1957 Act means invitees or licensees 'using the premises' – that is, anyone entering with permission or at the direct invitation of the occupier, or who enters under a right conferred by law (section 2 (6) 1957 Act). This covers most people lawfully on property, from relatives of the occupier, to tradesmen and the gas man. The 1984 Act applies to anyone else on the property, such as trespassers except for persons using the highway (section 1 (7) 1984 Act), and it does not apply 'to any person in respect of risks willingly accepted as his, by that person' (section 1(6)). Neither Act, therefore, applies to users of a footpath or bridleway across the occupier's land. Further, a person using a private right of way is not in general a visitor for the purposes of the 1957 Act: see *Holden* v. *White* [1982] QB 679.

10.4.3 Duty owed to visitors

The 1957 Act lays down that the occupier owes what it calls 'a common duty of care' to all visitors, except in so far as he can and does alter his obligation towards particular entrants by agreement. The duty is described as follows:

'The common duty of care is a duty to take such care as in all the circumstances of the case is reasonable to see that the visitor shall be reasonably safe in using the premises for the purposes for which he is invited or permitted by the occupier to be there.'

section 2 (2)

Elaborate protection of visitors is not expected. It is very much a duty of sweet reasonableness. If there are dangers on the land the visitor might get to, the occupier should either guard them or give sufficient warning of them. Although a warning will be taken into account in judging whether there is liability for a mishap (section 2(4)(a) 1957 Act), a warning will not always be sufficient protection. The Act admonishes occupiers to expect children to be less careful than adults (section 2(3)(a)).

The occupier is allowed to rely on visitors taking reasonable care of themselves, especially if the visitors are adults. Where someone is engaged to enter the property to exercise his calling (for example an agricultural worker, or a steeplejack) the occupier is not expected to protect him against the risks that go with the job (section 2(3)(b)). Further, if a visitor suffers damage due to faulty work by an independent contractor engaged by the occupier, it will normally be the contractor who is liable and not the occupier, provided it was reasonable for the occupier to assume a competent job had been done by him (section 2(4)(b)).

10.4.4 Agreements changing the duty

The 1957 Act allows the occupier to 'restrict, modify or exclude' the common duty of care, if he is free to do so. It is therefore open to him to say 'enter at your own risk' unless the Unfair Contract Terms Act 1977 applies. This 1977 Act applies 'only to business liability'. This will usually be where there is a contract entered into for a consideration. Where the 1977 Act applies the occupier cannot exclude by agreement liability for death or personal injury resulting from his negligence. Any agreement or notice to that effect would be a nullity. As regards a notice excluding liability for loss or damage (other than death or personal injury), this will be effective only if 'it should be fair and reasonable to allow reliance on it having regard to all the circumstances obtaining when the liability arose or (but for the notice) would have arisen' (section 11(3) 1977 Act). This leaves it uncertain and difficult to predict, as fair-and-reasonable tests always will.

The 1977 Act, however, proved too much of a deterrent to allowing recreational access to land where a business was carried on, such as agriculture or forestry. The 1984 Act therefore amended the 1977 Act. The effect is that occupiers may restrict, modify or exclude their duty of care towards anyone entering the land for recreational or educational purposes, unless granting the access 'falls within the business purpose of the occupier' (section 2 1984 Act). For example, someone running a safari park could not say 'enter at your own risk', but a farmer allowing orienteers onto his land, could.

10.4.5 *Duty owed to non-visitors*
Where the 1984 Act applies, namely when 'persons other than visitors', such as trespassers (see below) enter land, a duty is owed to the entrant if (1) the occupier knows of a danger on his land, or has reasonable grounds to believe it exists; (2) he knows, or has reasonable grounds to believe, the entrant is in, or may come into, the vicinity of the danger, and (3) 'the risk is one against which, in all the circumstances of the case, he may reasonably be expected to offer the other some protection' (section 1(3)). When those three conditions apply the duty owed 'is to take such care as is reasonable in the circumstances of the case to see that he (the entrant) does not suffer injury on the premises by reason of the danger concerned' (section 1(4)). In an appropriate case, the duty may be discharged by giving a warning or discouragement from incurring the risk (section 1(5)). 'Injury' here means death or physical or mental injury (section 1(9)) not loss or damage to property (section 1(8)).

It may be seen, the law is somewhat vague, turning on ordinary concepts of what is reasonable. It may well be that in most cases the occupier would not have to safeguard the *adult* trespasser beyond putting up notices warning of the more torrid and accessible dangers. If, however, *child* trespassers can be expected, greater safeguards must be taken to see that they do not fall foul of hazards. So in *Ratcliff* v. *McConnell* [1999] 1 WLR 670 a student who sustained tetraplegic injuries after diving into the shallow end of a college's open air swimming pool during restricted hours when the pool was closed for winter did not succeed in his claim partly because of the presence of a warning sign. However warnings will not be enough if, for example, there are gaps in protective fences that children can get through (see *Pannett* v. *P G Guinness & Co. Ltd* [1972] 2 QB 599 – child burned on demolition site). Before the 1984 Act, where a child got through a gap in a railway fence and was injured on the line, British Rail was held liable (*British Railways Board* v. *Herrington* [1972] AC 877) but not when an adult did so (*Titchener* v. *British Railways Board* [1983] 3 All ER 770). Occupiers should remember young people are attracted to certain hazardous places such as waters and derelict buildings, and in the eyes of the law this can make the occupier more likely to owe a duty of care.

10.4.6 *Employees*
The occupier has specific obligations for the wellbeing of his employees laid down in statutes and regulation, in particular the Health and Safety at Work, etc. Act 1974 and regulations' made under it. (See section 15 1974 Act for health and safety regulations generally; section 47 for extent of employer's civil liability.)

10.4.7 Landlords

Whilst a landlord who lets premises to a tenant is treated as parting with all control and is not an occupier under the 1957 Act, he can still be liable for the common parts, and for dangers he himself creates (*Anns* v. *Merton London Borough* [1978] AC 728). Further under the Defective Premises Act 1972 a special duty is placed on the landlord to take reasonable care to protect the tenant and 'all persons who might reasonably be expected to be affected' from defects if the landlord is responsible for maintenance and repairs under the tenancy, provided he knew or ought to have known of the defects (section 4 (1) and (2) 1972 Act).

10.5 Nuisance

The law of nuisance gives a right to an injured party to claim a remedy at law without having to show negligence. If the owner or occupier of land creates a nuisance it is no defence to prove that reasonable care was taken. Nuisance at law is what one might expect, namely causing a substantial nuisance to others. A nuisance may be public or private, as explained in sections 10.5.1, 10.5.2 and 10.5.3 below. Typical examples of nuisances found in the countryside are pungent smells, excessive noise from, for example, bird-scarers or saw mills, tree roots undermining neighbouring property and pollution of air, earth and water.

10.5.1 Public nuisance

A public nuisance is in essence a criminal act, not giving the individual a right of action except in special circumstances. It is an interference with the rights, comfort or convenience of the public or a section of the public. Normally it is for the Attorney General or the local authority to take proceedings to stop it on behalf of the public. It is open to an individual to take an action only if he suffers some particular damage over and above that suffered by the public generally. For example, obstructing the highway is a public nuisance, but an individual may sue if it obstructs access to his house, or prevents customers getting to his shop.

10.5.2 Statutory nuisance

Under Part III of the Environmental Protection Act 1990 local authorities have power to deal with 'statutory nuisances' as defined in section 79 by serving *abatement notices*. It is not uncommon for a draft notice to be served. In such situations, discussions with the environmental health officer can often resolve the situation without necessitating service of a formal notice. Where a formal notice is served, there is a right to appeal within 21 days (see the Statutory Nuisance (Appeals) Regulations 1995 (SI 1995/2644) for the full grounds of appeal). Otherwise it is an offence to fail to comply with such a notice, with a fine of up to £20 000 for statutory nuisances committed on industrial trade or business premises, and in respect of other land up to £5000 plus a maximum of £500 per day for each day the offence continues after conviction. The local authority may also abate the nuisance and recover its expenses. Statutory nuisances cover a wide variety of situations, such as smells, noise and threats to health. A local authority is under a duty to inspect its area for, and act if alerted to, any suspected public nuisances.

Local authorities have powers to deal with noise nuisance under Part III of the Control of Pollution Act 1974. Contravention of a noise abatement notice is an offence, but if the noise is caused in the course of trade or business it is a defence 'to prove that the best practical means have been used for preventing, of for counteracting the effect of, the noise'.

10.5.3 Private nuisance

10.5.3.1 Definition

A private nuisance is 'an unlawful interference with a person's use or enjoyment of land or some right over, or in connection with it' (*Read* v. *Lyons & Co. Ltd* [1945] KB 216). As with negligence, the tests the courts apply in deciding whether something is an actionable nuisance are tests of reasonableness and commonsense. A balance must be struck between the right of the occupier to do what he likes on his own property and the right of his neighbour not to have the enjoyment of his property interfered with (*Sedleigh-Denfield* v. *O'Callaghan* [1940] AC 880). People must expect farm land to be farmed and expect it to smell and sound like the countryside – but within reason.

10.5.3.2 The main rules

From the substantial body of decided cases the following main principles emerge.

(1) The nuisance must be unreasonable if it is to justify a legal remedy. The interference must be substantial: there must be too much of it. A farmer is allowed a manure heap, for example, but in *Bland* v. *Yates* (1914) 58 SJ 612 this caused an excessive number of flies to breed and infest neighbouring houses which was held to be 'a serious inconvenience and interference with the comfort of the occupiers of a dwelling house according to notions prevalent among reasonable English men and women'.

(2) The claimant must be in possession or occupation of the land, but persons with no proprietary interest cannot bring a claim. The essence of private nuisance is an interference with another in respect of his property – so that the wife of a tenant of land was once held to have no claim when vibrations from a neighbour's engine caused a water tank in their house to fall on her (*Malone* v. *Laskey* [1907] 2 KB 141, confirmed in *Hunter* v. *Canary Wharf Ltd* [1997] AC 655).

(3) The character of the locality must be considered where no physical damage is done. Where there is actual injury to property, it is no defence that the offending act was in keeping with the usual activities of the neighbourhood (*St Helens Smelting Co. Ltd* v. *Tipping* (1865) 11 HL Cas 642). However, where the complaint is interference with enjoyment of property with no physical injury done to it, the character of the neighbourhood must be considered 'and in particular with regard to trades usually carried on there' (*Polsue and Alfieri* v. *Rushmer* [1970] AC 121) to decide whether it is reasonable to allow the thing complained of to continue.

(4) It is no defence that the claimant moved to the nuisance. Provided the offending act would be a nuisance in any event, the claimant is not barred from a remedy even if he knew it existed before he went there. This is well illustrated by *Miller* v. *Jackson* [1977] QB 966. The plaintiff bought a house on the edge of a cricket field and was awarded £400 damages for the nuisance of sixes being hit into his garden (but in *Bolton* v. *Stone* [1951] AC 850 a passer-by struck by a cricket ball on the highway had no claim in nuisance or negligence).

(5) Special sensitivity does not give rise to a claim. If the complaint arises only because of the particularly sensitive nature of the complainant or his activities, no remedy may be claimed (*Robinson* v. *Kilvert* (1899) 41 ChD 88 where ordinary heating diminished the weight of brown paper sold by weight in what was an exceptionally delicate trade). On the other hand, if the nuisance would be actionable in any event, the claimant will not be barred from an action because of any oversensitiveness on his part (*McKinnon Industries Co. Ltd* v. *Walker* [1951] 3 DLR 577, 581), and a right to the high degree of light required for a greenhouse may be acquired by long enjoyment (*Allen* v. *Greenwood* [1979] 2 WLR 187).

(6) It is no defence that injury was due to a natural cause. When a landslip from Barrow Mump, a National Trust property, damaged a neighbouring house down the hill, the National Trust was held liable even though the slip was not its fault but a quirk of nature, as it was aware of the danger (*Leakey* v. *National Trust* [1980] 1 All ER 17). Because a positive duty is imposed to protect neighbours from natural dangers and this does not necessarily involve misfeasance, such liability is based on negligence. So in *Holbeck Hall Hotel* v. *Scarborough Borough Council* [2000] 2 All ER 705 CA the council was not held liable for a cliff slip that caused part of a hotel to collapse and the rest to be demolished because the danger and damage were far greater in extent than anything which was foreseen or foreseeable. These cases overrule *Giles* v. *Walker* (1890) 24 QBD 656, which had held there was no liability for thistles spreading from a

field onto neighbouring land. There will be no liability for an Act of God (see section 10.6.2 below).

10.5.3.3 Some types of nuisance

Livestock
In farming districts some noise and smell from animals is to be expected and must generally be tolerated. However, the smell from a large number of pigs kept in premises adjoining a village street was held a nuisance in *AG* v. *Squires* (1906) 5 LGR 99. Where extensive poultry farming was carried on in a rural area which was also residential, it was held by the court that a large number of cockerels crowing in the early morning was a nuisance (*Leeman* v. *Montagu* [1936] 2 All ER 1677).

Smells
A typical agricultural nuisance is smell from manure or effluent. In *Peaty* v. *Field* [1970] 2 All ER 895, manure spread on land near a housing estate was held to be a statutory nuisance under the Public Health Act 1936 (see below). In *Bone* v. *Seale* [1975] 1 All ER 797, neighbours brought a successful action for damages against a farmer because of the smell of pig manure, and the boiling of pig swill.

Wild animals
There used to be some support for the view that an owner or occupier could not be liable for the consequences of their presence of wild animals on their land, for instance see *Seligman* v. *Docker* [1948] 2 All ER 887, where the owner who kept pheasants on his land was held not liable to the adjoining owner when the pheasants increased naturally because it was a good season for them, nor was he required to keep the numbers down by shooting. However cases such as *Leakey* v. *National Trust* have eroded the principle that a person is not liable for natural events. This was confirmed in the specific instance of *ferae naturae* in *London Borough of Wandsworth* v. *Railtrack plc* [2001] 98 (37) LSG 38; [2002] Env LR 9. where the council established that Railtrack were liable in public and (potentially) private nuisance for the fouling and mess caused by pigeons congregating under a bridge.

In principle this decision could apply in the right circumstances to the effect of rabbits, pheasants and rats causing damage to adjoining land, but there are other procedures. Under statute an owner or occupier may be required to clear the land of rabbits, if a rabbit clearance order is made under the Pests Act 1954, or of a range of specified animals if a notice is served on him by the Secretary of State under section 98 of the Agriculture Act 1947.

Pheasants bred for sport can cause problems for adjoining farmers as pests. As between landlord and tenant, the farmer's remedy is compensation under the Agricultural Holdings Act 1986, section 20. There must be mutual reasonableness between the owner or tenant of sporting rights and the owner and occupier of the land: *Peech* v. *Best* [1931] KB 1 and *Pole* v. *Peake* [1998] EGCS 125. In the latter it was held that on the wording of an express reservation of shooting and rearing game, the rights could be exercised to a reasonable degree, notwithstanding interference with agricultural activities on the land.

Overhanging branches and encroaching tree roots
A branch overhanging another's property, or roots encroaching underground, can constitute a nuisance (this is not a trespass, as is commonly thought). No distinction is drawn between trees that are planted and those that are self-grown. In either case if roots or branches encroach and do damage an action for nuisance will lie. This can render the highways authority liable to an owner of a property adjoining the highway. The liability for damage is, however, dependent upon proof that the damage was reasonably foreseeable: *Solloway* v. *Hampshire County Council* (1981) 79 LGR 449. Nevertheless there is a right to abate the nuisance even though no action lies at law because of the absence of damage, and there is a specific power under section 154 of the Highways Act 1980 for the highways authority to cut back trees and foliage encroaching on highways.

As far as liability for a falling tree or branch is concerned, be it in respect of the highway or an adjoining property, liability is based on reasonable foreseeability of harm. So in *Cunliffe* v. *Bankes* [1945] 1 All ER 459, where a diseased tree fell on a passing motorist and killed him, the owner of the tree was not held liable because he had taken all reasonable steps to ascertain the condition of the tree and could not have realised it was likely to fall.

Straw and stubble burning
Section 152 of the Environmental Protection Act 1990 gives the Secretary of State power to restrict or ban the burning of straw and stubble burning. Unsurprisingly this led to substantial curtailment of this activity, currently by the Crop Residues (Burning) Regulations 1993 (SI 1993/1366). These prohibit the burning of straw and any other crop residue from cereal, field beans, oilseed rape and peas, save for educational purposes, disease control and disposal of broken bales. In such cases strict conditions are laid down as to the manner of burning, which also apply to the burning of linseed residues. Breach of the regulations is an offence with a maximum penalty of £5,000.

Weeds
Since the overruling of *Giles* v. *Walker* (see section 10.5.3.2, point (6) above) it is clear that the spread of weeds can give rise to liability. Nevertheless a plaintiff would have to show such degree of interference with his property that he cannot reasonably be expected to tolerate it.

There is also a power for the Secretary of State to serve notice in respect of notifiable disease under the Weeds Act 1959. Current policy seems to restrict this to benefit agricultural land only, although there is no legal foundation for this approach. Ragwort comes within the scope of this Act, and could potentially be the cause of an action in private nuisance.

10.5.3.4 Remedies

The usual remedy for a private nuisance is an action for damages and, if appropriate, an injunction, but self-help is allowed within limits.

Damages
Where loss or injury is proved the claimant is entitled to damages, whether the claim is for negligence or nuisance. It is often made up of two elements

(1) Loss or damage which can be quantified (such as loss of crop, or the cost of replacing broken window panes due to a flying cricket ball)
(2) Compensation for matters involving no pecuniary loss (such as for pain and suffering and loss of libido)

An important and often difficult rule to apply is that the damage claimed must not be too remote from the wrong complained of. The wrong must be the direct, though not necessarily the immediate, cause of the damage. The general test adopted is that any damage that the defendant could be reasonably expected to have foreseen as a result of the wrongdoing can be recovered (*The Wagon Mound (No. 2)* [1967] AC 667). In cases of strict liability (see section 10.6 below) the remoteness rule is less stringent.

Injunctions
An injunction is a court order to stop the activity resulting in the nuisance. To disregard an injunction is a contempt of court likely to be visited with condign punishment. Unlike damages, there is no entitlement to an injunction as of right. It is a discretionary remedy which the court will grant where justice so requires, and will generally only be granted where damages are an inadequate remedy in the circumstances. Injunctions may be granted not only to restrain existing wrongs but also threatened unlawfulness. An interim injunction will sometimes be made, especially in an emergency, to preserve the status quo until the case can be fully considered by the court. Injunctions are a flexible remedy, and can be granted on terms, such as the types of activity that can be granted and the times at which it may be conducted.

Self-help
The injured party is allowed to abate a nuisance, but it is a remedy to be resorted to with caution. It is only too easy for the wronged person to put himself in the wrong. If he resorts to self-help the method adopted to remove the nuisance must be a sound one doing no unnecessary harm or

damage and the abater must not cause a riot. Like Shylock, he must go thus far and no further. He must not enter the offender's land if it can be avoided and only after notice of entry if unavoidable. If these rules are followed, you may, for example, once you have given notice of your intention, cut off tree branches overhanging your land (*Lemmon* v. *Webb* [1896] AC 1) though the branches must be returned.

10.6 Strict liability: *Rylands* v. *Fletcher*

10.6.1 The rule

Where something potentially dangerous is kept on land the law allows the occupier hardly any excuse to avoid liability should it escape from the land. This rule of strict liability was expounded in the case of *Fletcher* v. *Rylands* (1866) LR 1 Ex 265 (confirmed in the House of Lords under its more familiar name of *Rylands* v. *Fletcher* (1868) LR 3 HL 330) in the famous words of Blackburn J

> 'The person who for his own purposes brings on his land and collects and keeps there anything likely to do mischief if it escapes, must keep it in at his peril, and, if he does not do so, is prima facie answerable for all the damage which is the natural consequence of its escape.'

In this case water escaped from a reservoir and flooded a mine – the court held there was strict liability. The rule only applies to non-natural uses of land – some special act of bringing something onto the land or accumulating something there. In practice it has proved increasingly difficult to draw the line between natural and non-natural user. Non-natural user would include the erection of a building or wall, and the working of a mine.

10.6.2 Defences

The law recognises only three possible defences if damage is caused by the escape

(1) that the escape was due to the claimants own fault; or
(2) it was an 'Act of God' – that is, a happening so unusual nobody could be expected to anticipate it, such as an earthquake; or
(3) it was caused by a stranger over whom the occupier had no control.

10.6.3 Duties under statute

There is technically a fourth defence, because the courts have held that where a body carries out duties required by statute, the statutory body is not liable for any damage caused in the absence of negligence, unless the statute expressly removes this exoneration. For example, a sewerage authority is not liable, in the absence of negligence, for damage caused by leakage from its sewers (*Smeaton* v. *Ilford Corporation* [1954] Ch 450).

However, the existence of this defence is now open to question following *Marcic* v. *Thames Water Utilities Ltd* [2001] 3 All ER, where it was held that a sewerage undertaker could be liable under the Human Rights Act 1998. At the time of writing this case was under appeal.

10.6.4 The Cambridge Water Company case

In *Cambridge Water Co. Ltd* v. *Eastern Counties Leather plc* [1994] 2 AC 264, the House of Lords gave the first serious consideration of *Rylands* v. *Fletcher* for some time. Unbeknown to the defendant, solvent used in leather manufacturing seeped through a concrete floor, into soil below and then into underground water used for the local public supply. However, the House of Lords refused to hold the defendant liable. Resisting the trend towards artificially restricting the meaning of non-natural user as a means of taking given instances outside strict liability, instead the House of Lords based its decision simply on the fact that events in question were unforeseeable. This case is to be seen as establishing a more conservative or restrictive approach towards the rule of strict liability.

10.7 Straying animals

10.7.1 Dangerous animals

Much of the law about escaping animals is now to be found in the Animals Act 1971. Section 2(1) imposes strict liability on the keeper of an animal of a 'dangerous species' for any damage caused by it. This is defined as animals of a species which is not commonly domesticated in the British Islands and when the animals are fully grown are likely to cause severe damage, or else any damage they may cause is likely to be severe (section 6(2)). The Dangerous Wild Animals Act 1976 provides for licensing the keeping of animals listed in it (SI 1984/1111 substituted a new list): see also the Zoo Licensing Act 1981. The Dangerous Dogs Act 1991 provides for pit bull terriers, tosas and other dangerous breeds added by regulation to be subject to restrictions and imposes criminal penalties for those who fail to observe the statutory restrictions.

10.7.2 Bulls

A bull is an animal domesticated in the British Isles and therefore not one of a dangerous species. The courts have the odd notion that bulls have at heart a kindly disposition and they must not be classed as likely to attack a man (see *Lathall* v. *Joyce & Sons* [1939] 3 All ER 854). Liability therefore turns, as with all other animals not of a dangerous species, on whether a particular animal was known to have a vicious tendency prior to an attack occurring.

For the special provisions relating to bulls in fields crossed by public rights of way see Chapter 6, section 6.3.11.

10.7.3 Animals not of a dangerous species

Under section 2(2) of the Animals Act 1971 the keeper is liable for damage done by these animals if three conditions are fulfilled:

(1) The damage done was of a kind the animal was likely to cause if not restrained, or which if caused by such an animal was likely to be severe
(2) The likelihood of that kind of damage is due to a characteristic in the particular animal which is not normally found in animals of the species, or is not normally found except at particular times or in particular circumstances
(3) Such characteristic was known to the keeper of the animal (or a servant in charge of the animal)

Thus, where a horse with a known nervous and unpredictable temperament injured a groom who was leading it into a horsebox, the groom recovered damages (*Wallace* v. *Newton* [1982] 2 All ER 106).

Where the keeper is the head of a household, knowledge of the characteristic by another keeper of the animal in the household under 16 years of age counts as his knowledge (Animals Act 1971, section 2(2)). The Guard Dogs Act 1975 provides for the control and licensing of guard dogs, but does not apply to dogs used to protect *agricultural* land nor to land within the curtilage of a dwellinghouse.

A magistrates' court may order the destruction of a dangerous dog (Dogs Act 1871) and, if it thinks fit, its owner to be disqualified from keeping a dog (Dangerous Dogs Act 1989).

10.7.4 Straying livestock

The Animals Act 1971 also imposes strict liability for damage done to land or property on land (though not for personal injuries) by trespassing livestock (section 4). 'Livestock' here means cattle, horses, asses, mules, hinnies, sheep, pigs, goats, poultry, and deer not in the wild state (section 11).

If the damage is the fault of the person suffering it there is no liability, but it must be remembered that in the absence of an express obligation to the contrary, it is the occupier's responsibility to keep in his own animals and not his neighbour's duty to fence them out.

The 1971 Act gives the occupier of land the right to detain and sell livestock straying on his land if nobody is in control of them. As with other remedies of self-help (see section 10.5.3.4 above) and because special rules have to be complied with (set out in section 7), it is a remedy to be resorted to with caution and usually best avoided. However this will not apply where the livestock has not 'strayed' onto the land as a result of the failure to fence in, but has been placed there. Indeed the owner of such animals may not be known. In such cases the only recourse is to sell under the procedure set out in sections 12 and 13 and Schedule 1 of the Torts (Interference with Goods) Act 1977 for the sale of uncollected goods.

10.7.5 Animals straying on to or off the highway

The Animals Act 1971 changed the law by creating a legal duty of care to prevent animals straying onto the highway (section 8(1)). This is not strict liability, so that if, for example, an accident was caused by a horse bounding out of a field onto a road, the person in charge would be liable if it was due to carelessness, but not if reasonable care had been taken to prevent it.

Further, if there is a right to place animals on common land, or on a town or village green or on land where fencing is not customary, the Act states it is not a breach of the duty of care to place the animals on such unfenced land (although once the animal strays beyond the point where the common land and highway co-exist, the owner will be criminally liable: see *Rees* v. *Morgan* (1976) 240 EG 787.

Nor is there strict liability when animals lawfully on the highway (e.g. livestock being moved) stray from it and cause damage (section 5(5) Animals Act 1971). In the absence of negligence there will be no liability for the damage. A famous incident is recorded in *Tillett* v. *Ward* (1882) 10 QBD 17, where a bull being driven to market along the street plunged into an ironmonger's shop and by the time it was extracted 45 minutes later, not a little havoc had been wrought. It was held the shopkeeper's claim failed as no negligence was proved. The law is the same today, but a commoner whose sheep strayed from unfenced common land via the highway into a garden was found liable for trespass (*Matthews* v. *Wicks*, *The Times*, 25 May 1987 (CA)).

10.7.6 Dogs worrying livestock

The keeper of a trespassing dog is liable if it kills or injures livestock (section 3, 1971 Act). The farmer is allowed to protect his livestock from a dog by shooting it, providing it is on his land (or his employer's) and the dog is in the act of worrying livestock, or is about to, and there are no other reasonable means of preventing or ending the attack. He may also shoot the dog if it has been worrying livestock, has not left the vicinity, is not under anyone's control and there is no practical means of ascertaining who the owner is (or he reasonably believes he cannot find out). He should notify the police within 48 hours of shooting the dog, because if he does so he will have a defence against a civil action if one is brought against him.

'Livestock' has the meaning given above with the addition of pheasants, partridges and grouse while in captivity (section 11, 1971 Act).

10.8 Trespass

Entry onto the land of another without consent or lawful authority is trespass. If a trespasser declines to leave the land on request, the occupier does not break the law if he physically ejects him, provided no more force is used than is reasonably necessary. It is generally advisable, however, not to resort to this remedy. Trespass is actionable *per se*, i.e. a trespasser may be sued even if no damage has been done. Normally there will be little purpose in taking proceedings unless significant damage has been done, or the offender is a persistent trespasser or further trespass is threatened. The court may be asked to award damages and/or issue an injunction to restrain trespass in the future (see, for example, *Patel* v. *W H Smith (Eziot) Ltd, The Times*, 16 February 1987).

Wilful trespass in numbers has presented increasing problems for occupiers on a number of fronts, from encamped travellers to raves to protests. Trespass is not a crime itself, but the Criminal Justice and Public Order Act 1994 strengthened police powers to deal with this. Section 61 of the Criminal Justice and Public Order Act 1994 gives certain powers to the police to order encamped trespassers to leave land. A person who fails to leave land when directed to do so, commits an offence. The new provision applies where two or more persons have entered land as trespassers, and have the common purpose of residing there. Reasonable steps must have been taken by or on behalf of the occupier to ask them to leave, and the trespassers must have done damage to property on the land or used threatening behaviour, or have brought 12 or more vehicles on to the land. A trespasser who fails to leave the land as soon as reasonably practicable or who re-enters as a trespasser within three months of a direction to leave being given, commits an offence. Land is defined in the section to include farm buildings and scheduled monuments.

The Civil Procedure Rules (formerly the Rules of the Supreme Court) have been amended to enable occupiers suffering from wilful trespass in numbers to get eviction orders quickly. Where trespassers squat on common land with no registered owner, the local authority can obtain a possession order under section 9 of the Commons Registration Act 1965 (see *R* v. *Teignbridge District Council ex parte Street, The Times*, 9 October 1989).

There are also special powers under section 63 of CJPOA 1994 to deal with 'raves'. A rave is defined as an unlicensed gathering of 100 or more persons (whether or not trespassers) on land in the open air (or partly so) at which amplified music is played during the night and by reason of its loudness and duration and the time at which it is played, is likely to cause annoyance to local inhabitants. Where a police officer believes that two or more persons are preparing to hold a rave, or ten or more are attending such an event or waiting for it to commence, they may give a direction for those persons attending to leave the land and remove any vehicle or their property they have with them on the land. It is then an offence for any such person to fail to leave the land as soon as reasonably practicable or to return within seven days. There are ancillary powers for the police to seize equipment and to direct people away from raves within a five mile radius.

Section 68 of CJPOA 1994 introduces a new offence of aggravated trespass. This applies where a person is on open land and does anything in relation to any lawful activity which other persons are engaged in (or about to be) which intimidates any of those persons so as to deter them from engaging in that activity, obstructs that activity or disrupts that activity. This is an arrestable offence. There are also provisions relating to trespassory assemblies (section 70).

Chapter 11

Water and Watercourses

Abbreviations in this chapter
'WRA 1991'   Water Resources Act 1991
'EA 1995'   Environment Act 1995
'IDB'   internal drainage board
'LDA 1991'   Land Drainage Act 1991
'Secretary of State' Secretary of State for Environment, Food and Rural Affairs, or the National Assembly for Wales where the context requires

11.1 Responsibility for water management

11.1.1 The Environment Agency

Prior to the Water Act 1989 the water authorities were responsible for a wide range of water related matters as well as water supply and sewage disposal. The Water Act 1989 revolutionised this, privatising the ten regional water authorities in England and Wales and creating the National Rivers Authority. In 1991 there was a consolidation process in which most of the then existing legislation was brought together in the Water Resources Act, the Water Industry Act, the Statutory Water Companies Act and the Land Drainage Act 1991. This framework retains the separation of the regulatory functions from the water supply and sewerage responsibilities, which was established in 1989.

The Environment Agency was established on 1 April 1996 under the EA 1995 to provide integrated pollution control and management of rivers, basins and the water environment in England and Wales. It has inherited those functions of the National Rivers Authority relating to fisheries, flood defences, pollution control, water resources, navigation, conservation and recreation. The headquarters is in Bristol, and there are eight catchment based regional offices. The agency board may consist of least 8 but not more than 15 members, appointed by the Secretary of State, who also appoints the chairman. One member may also be appointed by the National Assembly for Wales. Within its various functions, that of flood defence is unique, being discharged through 10 executive *regional flood defence committees* again catchment based, with in some regions *local flood defence committees* being in place.

In general the Environment Agency may regulate its own procedure with the exception that its flood defence functions must be carried out through regional flood defence committees and may not be exercised by any other body. The Environment Agency must establish and maintain *environment protection advisory committees* for the different regions of England and Wales, and consult them in the discharge of its functions.

11.1.2 Water undertakers

Prior to 1 September 1989 water and sewerage services were generally provided by the public water authorities established under the Water Act 1973. The Water Act 1989 transferred the functions of each water authority to successor public limited companies, which provided water supply and sewerage services as commercial enterprises. The governing legislation is now contained in the Water Industry Act 1991. The Director-General of Water Services is responsible for the economic regulation of the water and sewerage industry, including the appointment of water and sewerage undertakers for each given area. Through the Office of Water Services (Ofwat) he is concerned with the protection of customers' interests as regards charges and levels of service. The undertakers are also regulated by the Secretary of State at DEFRA in England, and the National Assembly in Wales.

11.2 Land acquisition

The Environment Agency and the water undertakers have power to acquire land and rights over land by agreement or compulsorily for the performance of their functions under the Water Resources Act 1991 section 154 and the Water Industry Act 1991 section 155. The procedures and principles for compulsory purchase are generally as explained in Chapter 8 and for the acquisition of wayleaves for water pipelines and sewers in private land refer to section 9.7, Chapter 9.

11.3 Riparian rights

11.3.1 Meaning

Land alongside a river or other water is known as *riparian* land and the owners have certain rights at common law deriving from their ownership of riparian land. Riparian fishing rights are explained in Chapter 12 at section 12.6.2.

11.3.2 Rights to use water

The water flowing in a river is ownerless. The riparian owner, however, has a right at common law to have the water flowing through or past his land undiminished in quality or quantity (*Scott-Whitehead* v. *National Coal Board* (1987) 53 P & CR 263) and has the right to take and use the water, so long as it is consistent with the rights of the riparian owners upstream and downstream of him. This means he can take the water for his own domestic purposes and for his livestock, even if by so doing he leaves insufficient for lower riparian owners, but if he takes it for other purposes and thereby leaves lower owners short to their detriment, the lower owners can claim damages and an injunction. Taking water for spray irrigation is not a riparian right (*Rugby Joint Water Board* v. *Walters* [1966] 3 All ER 497) but taking it for ordinary purposes in livestock buildings, such as washing down, would be within the riparian right.

11.3.3 Subterranean water
The riparian right to receive water as described above applies to underground water only if it flows in a defined channel. In the more usual case of underground water percolating which is not in a defined channel, there is no neighbourly obligation at law: (*Chasemore* v. *Richards* (1859) 7 HL Cas 349). There is no common law remedy against an abstractor who dries up such groundwater whatever the purpose of the abstraction.

11.3.4 Statutory curtailment of rights
The common law riparian rights and obligations are subject to the laws regulating abstractions in the Water Resources Act 1991.

11.4 Water abstraction

11.4.1 The general rule
The general rule is that water may not be abstracted from a 'source of supply' unless it is permitted by licence granted under Part II of the WRA 1991. Abstraction of water from a source of supply means the doing of anything as a result of which water is removed from a source of supply, whether temporarily or permanently, and whether it is transferred to another source. There has to be a positive act, and mere inducement of a gravitational flow does not constitute an abstraction: *British Waterways Board* v. *Anglian Water Authority*, *The Times*, 23 April 1991. There are exceptions to the licensing requirements, listed in section 11.4.3 below, which preserve to some extent common law riparian rights to take water. The granting of abstraction licences is a function of the Environment Agency.

11.4.2 Meaning of 'source of supply'
The definition of 'source of supply' can be found in section 221 of the WRA 1991. In short it means any inland water, meaning any natural or artificial river, stream or watercourse whether tidal or non-tidal; any lake, pond or reservoir unless it does not discharge into an outside water; a channel, creek, bay, estuary or arm of the sea; or water in any underground strata, or well, or borehole or similar work, including an adit.

11.4.3 Exemptions from licensing
No abstraction licence is needed in the following cases.

(1) Abstractions up to 5 m^3 of water, if they are not part of continuous, or a series of, operations (section 27(1) WRA 1991). A similar type of abstraction of up to 20 m^3 is allowed, with the consent of the EA (section 27(2)).

(2) Abstractions (not of underground water, for which see (3) below) for use of the riparian occupier on the riparian holding for the domestic purposes of the occupier's household and agricultural purposes other than spray irrigation (section 27(3) and (4) WRA 1991), limited to 20 m^3 in any 24-hour period. The exception for the occupier's household does not extend to other dwellings on the farm or estate, such as workers' cottages.

(3) Abstractions of underground water by an individual for the domestic purposes of the occupier's household (though not for agricultural purposes) limited to 20 m^3 in any 24-hour period (section 27(5) WRA 1991). He may also carry out works to obtain the water (section 27(7)).

(4) Abstractions in the course of, or resulting from, land drainage operations (section 29(1) WRA 1991).

(5) Abstractions used in connection with de-watering operations (section 29(2)).

(6) Abstractions for fire fighting; for consented test boring; or under an order of the Secretary of State following an application by a 'relevant authority', e.g. a harbour authority (sections 32(2), 32(3) and (4), and 33 WRA 1991).

11.4.4 Abstraction licences
Where a licence is required, it is an offence, punishable by a fine, to abstract the water, or to 'cause or permit any other person' to do so, or to construct or extend any well, borehole or other work for abstraction without a licence, or to contravene the conditions of a licence (section 24 WRA 1991).

There are two kinds of licences, namely licences issued at the discretion of the Environment Agency and licences of right, to which certain qualifying persons were entitled. However, the deadline for applying for licences of right passed on 30 June 1965 (see section 65 and Schedule 7 WRA 1991).

Only the occupier or prospective occupier of the land on which the abstraction point is located may apply (see section 35 WRA). A person can also apply to abstract water from a source of which he is not the owner or occupier, but he must demonstrate that he has permission to access that source.

Every licence must state, among other things, the quantity of water that may be abstracted from the source of supply, over what period, by what means, where it may be used and for what purpose, and how it is to be measured (section 46 WRA 1991).

Charges are made for licences, based on a scheme made under section 41(1)(a) and (8) of the EA 1995. These can be an administrative fee for processing the application, and then a single sum for the period of the licence, or more normally an annual quantity charge. Whichever is levied, these will depend on the amount of water authorised to be abstracted, although for some uses, notably spray irrigation, the charge can be based in part on licenced quantity and part on water used.

11.4.5 Application for licence

Application must be made to the Environment Agency, accompanied by a notice in a prescribed form and evidence must be produced showing the following:

(1) That the notice has been published in the *London Gazette* and in two successive weeks in at least one newspaper circulating in the locality.
(2) If the abstraction would be from an 'inland water' (that is, not an underground source), that the notice has been served on any local water undertaker, and any navigation, harbour and conservancy authorities concerned and on any internal drainage board.
(3) If the abstraction would be from underground strata, that the notice has been served on any local water undertaker.

The notice must also state where the application with its maps and plans may be inspected. The public must be allowed to inspect them free of charge at any reasonable hour during a period of 28 days after publication of the notice and any person may make objections in writing to the Environment Agency during that 28-day period. The Environment Agency may dispense with these requirements if the application is for less than $20\,m^3$ per 24 hours (section 37 WRA 1991).

11.4.6 Determination by the Environment Agency

The Environment Agency must make its decision within three months of receipt of the application (a deferral can be requested in proper circumstances), or notify the applicant that the application has been called in by the Secretary of State for local inquiry. It must have regard to any representations made as a result of the publicity and to the requirements of the applicant so far as they seem reasonable. The Environment Agency must also have regard to other factors, including the character of any inland water affected and minimum acceptable flows and the requirements of existing lawful uses of the water, whether for agriculture, industry, water supply or other uses and the requirements of land drainage, navigation and fisheries (sections 38 and 40 WRA 1991).

In particular, the Environment Agency must not grant a licence for an abstraction which would derogate from the right of existing licence holders, or from rights to abstract which are exempt from licensing (section 39 WRA 1991). Should the Environment Agency derogate from such rights it will not invalidate the licence but the Environment Agency may be sued for damages for breach of statutory duty (section 60).

11.4.7 Appeals and public inquiries
If the abstraction licence is refused or the applicant is dissatisfied with the conditions in a licence, he may appeal to the Secretary of State within one month. All objectors must have a copy of the appeal served on them by the Environment Agency and they may make further representations. The Secretary of State may hold a public inquiry (and must do so if the applicant or the Environment Agency so requests), or give the applicant or the Environment Agency an opportunity for a hearing before a person appointed by him. The Secretary of State's decision on the appeal shall be final (sections 43–45 WRA 1991).

11.4.8 Protective effect of licence
If any action is brought against a person in respect of an abstraction it shall be a defence to prove the abstraction was in accordance with an abstraction licence. The licence does not, however, protect the holder from an action for negligence or breach of contract (section 48 WRA 1991). Apart from this, the only remedy of a third party will be against the Environment Agency for breach of statutory duty, if this can be shown. Abstraction without a licence, when one is required, does not of itself give a neighbour a legal claim against the abstractor, or allow the neighbour to interfere with an easement to abstract water (*Cargill* v. *Gotts* [1981] 1 All ER 682).

11.4.9 Changes in occupation
The abstraction licence in respect of a holding will be in the name of the occupier. If he dies or there is a change in occupation for any other reason, the new occupier must notify the Environment Agency within 15 months and he will then become the licence holder. Otherwise the licence will become invalid (section 49 WRA 1991).

11.4.10 Variation and revocation of licences
An application for a variation in the licence by the abstractor is treated as a fresh application for procedural purposes, although if the variation is to reduce the quantity of water to be abstracted, then the charge and requirement for advertisement are usually dispensed with (section 51 WRA 1991).

The Environment Agency can propose a variation (sections 52–53 WRA 1991), and the owner of fishing rights can apply to the Secretary of State for a variation or revocation (section 55).

11.4.11 Minimum acceptable flows
The Water Resources Act 1963 imposed a statutory duty on the water authorities and their predecessors to fix 'minimum acceptable flows' for each inland water which could be subject to licensing, so that there might be a yardstick for the proper determination of abstraction licence applications (section 19 WRA 1991). There was a total dereliction of this duty, perhaps due to the difficulty of doing this in practice. Not a single minimum acceptable flow was set. This changed with the 1989 Act, and the present position is that the Environment Agency is empowered to submit proposals for the Secretary of State to do so, but no duty is imposed to determine minimum acceptable flows. The Secretary of State can also direct the Environment Agency to prepare proposals for particular waters. Where minimum acceptable flows are fixed, the Environment Agency can vary or revoke abstraction licences accordingly.

11.4.12 Spray irrigation
Where a licence authorises spray irrigation, the licence may impose temporary restrictions on the quantity of water that may be abstracted for this purpose in times of exceptional shortage of rain, or other emergencies. The Environment Agency in so doing must be even-handed between abstractors from the same source of supply who are not far distant from each other (section 57 WRA 1991).

11.4.13 Drought orders
The Secretary of State has wide powers to protect water sources and water supplies by making drought orders under the WRA in times of shortage of water (sections 73–81 WRA 1991).

11.5 Impounding water

11.5.1 Impounding works
Abstraction charges are normally less if water is taken only in the winter when it is normally plentiful. Landowners are thereby encouraged to construct reservoirs. Before building a reservoir, or say, a fishing lake, the Environment Agency must be consulted because 'impounding works' may not be carried out without a licence (section 25 WRA 1991). Contravention is an offence punishable by a fine (section 25(3)).

'Impounding works' means any dam, weir or other works in an inland water whereby water may be impounded and any works for diverting the flow of an inland water in connection with the construction or alteration of such works (section 25(8) WRA 1991). In most cases such works will also require planning permission.

11.5.2 Impounding licence

The procedures for obtaining impounding licences and for appeals against the determination of the Environment Agency are similar to those for licences (sections 11.4.4–7 above).

11.6 Stocking fishing lakes

An owner who wishes to stock a lake or reservoir with fish should note that he will need the prior consent of the Environment Agency before introducing any fish or spawn into the water. Contravention is an offence punishable by a fine (Salmon and Freshwater Fisheries Act 1975, section 30 and Schedule 4). Further, the Wildlife and Countryside Act 1981 makes it an offence to release, or allow to escape, into the wild, certain non-indigenous creatures, including zander, wels and pumpkinseed fish (section 14 and Schedule 9).

11.7 Land drainage and flood defence

11.7.1 Regulatory bodies

The bodies involved in land drainage and flood defence are the Environment Agency, IDBs and local authorities (predominantly district councils). The responsibility for official land drainage and flood defence falls to the Environment Agency and the IDBs, and with the former being governed by the WRA 1991 and the later by the LDA 1991.

All involved bodies have powers to regulate activities in and alongside waterways which could adversely impact upon the flow of water. This would include the installation of culverts, alterations of weirs, etc. Under byelaws they are also able to control a strip alongside waterways, generally 7–9 metres, so that access to the waterway for maintenance purposes was not impeded.

11.7.2 Main rivers

The Environment Agency's land drainage and flood defence function is principally to ensure that watercourses designated as *main rivers* convey water effectively. It is carried out by the Environment Agency's flood defence committees. A 'main river' is one designated as such by the Secretary of State (sections 113(1) and 194 of the WRA 1991). The Environment Agency has no power to do land drainage works to other watercourses. The 'main river' in any area is a watercourse shown on the main river map prepared by the Secretary of State and held by the Environment Agency for public inspection (section 193).

An owner or occupier cannot compel the Environment Agency to undertake drainage works to a main river, but the Secretary of State can direct the Environment Agency to carry out its land drainage (and other) functions (section 40 EA 1995). Further, the Secretary of State can hold a public inquiry to see whether the Environment Agency has failed to perform any of its functions and if it is found to be in default can take steps to see the default is remedied (section 53 EA 1995; section 69(1) LDA 1991).

The Environment Agency is also responsible for fluvial, tidal and coastal flood defences and the issuing of flood warnings.

11.7.3 Internal drainage boards

Internal drainage boards administer internal drainage districts which are areas which do not flood or derive benefit from land drainage operations. The boundaries which are on a Ministry sealed plan are based generally to include areas which are up to 2.46 metres (8 feet) above the highest known flood for non-tidal situations or 1.54 metres (5 feet) above the ordinary spring tide for tidal situations. Areas above these levels but which would otherwise be cut off due to flooding can be included in the drainage district.

An IDB has power to carry out works of maintenance or improvement to any watercourse in its district, and has powers of compulsory acquisition (section 62 LDA 1991). Again, an IDB cannot be compelled by an owner or occupier to do drainage works.

11.7.4 Local authorities

Local authorities confine their activities to the flood protection of property in urban areas and will seldom, if ever, attend to rural watercourses other than where they have been 'awarded' to the authority under the Inclosure Acts. They can serve notice on a riparian owner to maintain a watercourse other than the main river or within an internal drainage district.

11.7.5 Liability of drainage authorities

The Environment Agency and IDBs are not legally liable for damages for failing to carry out drainage works. On the other hand if any injury is caused to any person by works carried out by a drainage authority they 'shall be liable to make full compensation' (section 177 and para 5(1) of Schedule 21 WRA 1991 and section 14(5) of LDA 1991). 'Injury' has its usual legal meaning of loss or damage, and can include fishing and other rights: *Burgess* v. *Gwynedd River Authority* (1972) 24 P & CR 150.

The statutory compensation is not applicable where injury arises out of the exercise of the body's statutory powers, but where the injury is suffered as a result of the negligent act occurring on the course of the statutory power, a civil action may be brought: *Marriage* v. *East Norfolk Rivers Catchment Board* [1949] 2 All ER 1021. If there is any unnecessary damage or interference with rights (such as fishing rights) it will be an actionable wrong (*Welsh National Water Development Authority* v. *Burgess* (1974) RVR 395 (CA)).

Where a watercourse is widened or dredged the drainage authority may 'appropriate and dispose of any matter removed' without payment for it (section 15 LDA 1991). The Act says matter may be deposited on the banks over such width as to enable it to be removed and disposed of in one mechanical operation, in which case the authority *may* pay compensation 'if they think fit' (section 15(1)(b)). The Lands Tribunal, awarding damages and interest in such a case, held that timber and fruit trees removed for widening a watercourse were not 'matter' and must be paid for, and although no payment was to be made for the land taken by the widening, as it was 'matter', compensation was payable to the farmer for the loss of use of the land removed (*Pattinson* v. *Finningley* IDB (1971) 22 P&CR 929).

11.7.6 Drainage revenue

The Environment Agency raises revenue for flood defence through levies on constituent county councils and unitary authorities and IDBs. It can also place a general drainage charge on all agricultural land outside the internal drainage districts, but this is currently applied only in the Anglian region. It can also secure contributions from developers and others towards the cost of undertaking works necessary to ensure that there is no adverse flood risk impact arising from proposals.

For certain approved works, the Environment Agency receives a grant from DEFRA.

Within its district an IDB has power to raise revenue by levying drainage rates on agricultural land and buildings (see Chapter II of Part IV LDA 1991) which are payable by the occupier (section 40) – a tenant can no longer deduct the rate from the rent. They also raise income through a special levy served on constituent district councils and unitary authorities payable in respect to all the non-agricultural properties and land benefiting from the IDB work.

11.7.7 Getting neighbours to do drainage works

A landowner or occupier can apply to the Agricultural Land Tribunal for an order requiring another owner or occupier to cleanse, or restore a ditch to proper order. The Tribunal will consider whether work is necessary to the ditch to enable the applicant to drain or improve his land and may, after hearing the parties, make such order as it sees fit. If it orders a person through whose land the ditch passes or abuts to do work to the ditch and he defaults, after three months the Department for Environment, Food and Rural Affairs or a drainage body may carry out the work and recover the cost from the defaulter (sections 28–29 LDA 1991). The Tribunal also has power, on the application of an owner or occupier, to authorise him to go onto another's land to carry out drainage works or to alter a ditch (section 30 LDA 1991).

In either case, the authorised person may then, after giving at least seven days' notice, enter with men and equipment to do the necessary work. If the land is unoccupied, it must be left secure against trespass, and in any case, if unnecessary damage is done he will be liable to pay compensation (section 29(3)–(7) LDA 1991).

A landowner or occupier may refer the matter to the Environment Agency, IDB or local authority under section 25 of the LDA where similar powers exist.

There is also power for the Secretary of State to authorise an owner or occupier to enter on the land of others who object to desirable drainage works, to do the works. If there is an objection to the application, the Secretary of State must first hold a public inquiry (section 22 LDA 1991).

The Environment Agency or a drainage authority has power of its own volition, by notice, to require landowners and occupiers to remedy the condition of watercourses (section 33 LDA 1991 and section 107 WRA 1991).

Chapter 12
Sporting Matters

Abbreviations in this chapter:

'1831 Act'	Game Act 1831
'1860 Act'	Game Licences Act 1860
'1880 Act'	Ground Game Act 1880
'1981 Act'	Wildlife and Countryside Act 1981
'1968 Act'	Firearms Act 1968
'1975 Act'	Salmon and Freshwater Fisheries Act 1975
'1988 Act'	Firearms (Amendment) Act 1988
'the Secretary of State'	Secretary of State for Environment, Food and Rural Affairs, or the National Assembly for Wales where the context requires

12.1 Sporting rights

12.1.1 Sporting rights are property
Sporting rights may exist as a legal interest in land. From this simple point of jurisprudence much flows. The rights may be owned together with or separately from the land. They may be sold or leased or given away. They may be defended against trespass or interference (*Mason* v. *Clarke* [1955] 1 All ER 914) and, on a compulsory acquisition, must be paid for. In exercising the rights the owner, whether or not on his own land, must comply with the laws concerning close seasons, protected species, cruelty to animals, licensing and firearms.

12.1.2 Who may exercise sporting rights?
To carry out sporting activities lawfully a person must own the sporting rights or have obtained the necessary right or permission from the owner by lease or licence. In certain places the public may exercise public fishing rights.

To ascertain who owns the sporting rights over any land is a matter of tracing the history of ownership. If they have never been disposed of separately from the land, the landowner will own them.

12.1.3 Farm tenants
When land is leased to a tenant, the tenant obtains a tenancy of the sporting rights as well, unless the landlord reserves them (*Pochin* v. *Smith* [1887] 52 JP 4) or they were owned by someone else in the first place, with one exception – he may not kill ground game at night with a gun: see section 6 of the 1880 Act. It is not common in a farming tenancy for the landlord to reserve the sporting rights to himself.

The tenant of an agricultural holding can claim compensation from his landlord if his crops are damaged by any wild animals or birds the landlord (or anyone claiming under him) has the right to kill, provided the tenant has not permission in writing to kill them (section 20 Agricultural Holdings Act 1986). The compensation can only be claimed if notice in writing is given to the landlord within one month after the tenant first became, or ought reasonably to have become, aware of the damage and a reasonable opportunity is given the landlord to inspect the damage before the crop reaping begins (or if the damage is to a gathered crop, before it is removed). Written particulars of the claim must also be given 'within one month after the expiry of the year in respect of which the claim is made', the 'year' ending on 29 September, unless otherwise agreed. In cases where the landlord does not have the sporting rights, he is entitled to be indemnified by whoever has them against claims for game damage. See also Chapter 10 at 10.5.3.3 for possible liability in game damage nuisance.

12.1.4 Occupier's right to ground game

No matter who may have the sporting rights the occupier of the land has the right conferred by the Ground Game Act 1880 to take and kill ground game (namely hares and rabbits) so that he may protect his crops. Any agreement to relinquish this right will be null and void. If someone else has the right to ground game (for example the landlord or the shooting tenant) it will be exercised along with the occupier of the land.

Where the occupier does not have the shooting rights, his right to take ground game under the 1880 Act is subject to restrictions. Only the occupier and one other person, who must be authorised by the occupier in writing, may use firearms to kill ground game under the Act. More than one other may be authorised (in writing) by him to take ground game by other methods (for example, with snares or ferrets), but a person can only be authorised by the occupier if he is

(1) A member of his household resident on the land
(2) An employee in ordinary service on the land (for example a farm worker)
(3) Any one other person bona fide employed by the occupier for reward to destroy ground game

Any such authorised person must, if required by anyone having a concurrent right to ground game, produce his written authority.

A person with a right of common over common land does not have the benefit of an occupier under the Act, nor does a person who occupies 'for the purpose of grazing or pasturage of sheep, cattle or horses for not more than nine months' (section 1). The occupier and one other person authorised by him may now shoot ground game at night with the written authority of any one other person so entitled (Schedule 7, 1981 Act).

12.1.5 Rating
The rating of sporting rights was abolished as from 1 April 1997.

12.1.6 No public shooting right
A public right of shooting is not known in English law.

There is a presumption that the Crown owns the foreshore, and the burden of proof to the contrary is on a claimant. There are only two public rights over Crown foreshore, namely fishing and navigation and there are no public rights over private foreshore. However, shooting is tolerated by the Crown over much foreshore, provided others are not endangered, but the wildfowler cannot claim to shoot as a public right (*Fitzhardinge (Lord) v. Purcell* [1908] 2 Ch 139).

Again there is no public right to shoot on roads and rights of way (*R v. Pratt* (1855) 4 E&B 860) the public right being limited to 'passing and repassing'. The soil of a highway, however, belongs to the adjoining landowner, so it is permissible for the property owner to shoot from the highway provided an offence is not committed against the Highways Act 1980, (section 161 (as amended)), which forbids the discharge of a firearm within 50 feet of the centre of the highway comprising a carriageway if in consequence 'a user of the highway is injured, interrupted or endangered'.

12.2 Protection of wildlife

12.2.1 Legislation
The protection of birds and animals in the wild is now dealt with largely by the Wildlife and Countryside Act 1981 and various Acts dealing with game and deer.

12.2.2 Birds
All wild birds, their eggs and nests are protected in Great Britain at all times, except in so far as the 1981 Act (and the game laws, see section 12.3 below) allow specified species to be killed or taken (sections 1–8 and Schedules 1 and 2 1981 Act). It is generally a defence to a charge if a bird is killed of necessity to prevent serious damage to livestock, foodstuffs for livestock, crops, vegetables, fruit, growing timber or fisheries, or to prevent spread of disease (section 4(3) 1981 Act). Other exceptions include certain official action (section 4(1)) and acts of mercy to disabled birds (section 4(2)).

The general scheme for the protection of birds is to create offences of killing, injuring or taking birds, and of taking, damaging or destroying nests or eggs, and of having possession of birds, nests or eggs. There are also offences concerned with selling them but two general licences came into effect on 1 January 1995 regarding the sale of certain feathers and dead birds (WLF 100092/-96). Rare birds listed in Part I of Schedule 1 of the 1981 Act are protected by special penalties at all times, those listed in

Part II of Schedule 1 are protected by special penalties in the close season for the bird in question. The close seasons are set out in section 2(4), but may be varied (section 2(5)). There are also special penalties for disturbing Schedule 1 birds while they are nesting, building nests or have dependent young (section 1 (5)). Birds listed in Part I of Schedule 2 may be killed or taken only outside the close season. Those in Part II of Schedule 2 may be killed or taken by authorised persons at all times. 'Authorised persons' include the landowner or occupier (includes a person with the sporting rights) (section 27 (1)). The Schedules may be varied by order.

Provision is made for temporary orders protecting birds that could otherwise be shot, for up to 14 days (section 2(6) and (7)) and for specifying areas of special protection (section 3). Schedule 2 birds may not be killed on Sundays or Christmas Day in Scotland nor on Sundays in England and Wales where orders are made prohibiting it (section 2(3)).

12.2.3 Animals

The 1981 Act affords protection to certain species of wild animals (see sections 9–12 and Schedules 5–7) none of which would normally be the prey of legitimate sportsmen. Rabbits and hares are dealt with by the game laws (see section 12.3 below). The Deer Act 1991 provides a close season for four species of deer, and prohibits the use of certain weapons (see section 12.5 below). It also creates offences against poaching and in relation to sales of venison.

There is specific protection for badgers in the Protection of Badgers Act 1992 which creates various offences relating to the killing and infliction of cruelty on badgers, and interference with a badgers' sett or disturbing badgers. There is provision for licences (section 10) to kill badgers or interfere with a sett if it is necessary to prevent serious damage to land, crops, poultry and other property.

The Wild Mammals (Protection) Act 1996 creates a general offence of inflicting cruelty on any wild animal with intent. There are limited defences, such as mercy killings. The Protection of Animals Act 1911 contains a number of offences against animals for cruelty. The provisions of the Act apply to all domestic and captive animals, as defined therein. Therefore this will include wild animals when in captivity or confinement (such as pheasants in pens). The Protection of Animals (Amendment) Act 2000 applies to the same animals as the 1911 Act so long as they are kept for commercial purposes. Amongst other things it provides for the care, disposal or slaughter of animals in the interests of animals' welfare, and for various persons including a person who is approved by the Department for the Environment, Food and Rural Affairs or the Welsh Assembly to bring proceedings under section 1 of the 1911 Act.

12.3 Game laws

12.3.1 Definition of game

Unfortunately 'game' means something different in nearly every Act referring to it. The definition must be checked in each case. The Court of Appeal held in *Inglewood Investment Co. v. Forestry Commission* [1989] 1 All ER 1 that in a reservation of 'game woodcock snipe...' etc., 'game' did not include deer on the wording of the lease.

There is no close season for killing rabbits and hares, but game, including hares, may not be killed on Sundays or Christmas Day (Game Act 1831, section 3).

12.3.2 Game licences

It is an offence to pursue, take or kill any game, or to assist in doing so, or to use any dog, gun, net or other engine or instrument for the purpose (including a snare: *Allen v. Thompson* (1870) LR 5 QB 336) without a game licence. 'Game' here means pheasants, partridges, grouse, black-game, moorgame, hares, woodcock, snipe, rabbits and deer (section 23 1831 Act and section 4 Game Licences Act 1860).

None of the following needs a game licence: beaters or persons assisting a licence holder, provided they are not carrying guns; persons killing ground game under the Ground Game Act 1880 (see section 12.1.4 above); the owner or occupier killing hares or deer on enclosed lands; or taking rabbits at a warren on enclosed lands; persons coursing or hunting hares with dogs or hunting deer with hounds; anyone required by the Secretary of State to kill game by order under the Agriculture Act 1947.

Anyone paying the licence duty is entitled to a game licence as of right. It is obtainable from Post Offices. The licence comes into force from the moment it is issued and it cannot be retrospective. It is not transferable. In practice these requirements seem rarely observed, although this remains a criminal offence.

12.3.3 Dealing in game

Anyone dealing in game must have two licences, one from the local authority and the other an excise licence: section 18 1831 Act and section 14 1860 Act. This does not mean the owner of shooting rights holding a game licence cannot sell the bag, lawfully shot, to the butcher. He may do so provided the butcher is a licensed dealer in game.

12.4 Poaching of game

12.4.1 Statutory offences

The laws against poaching are complex, being spread around several pieces of legislation, without any rationalisation and consolidation. They are still mainly found in the venerable Night Poaching Acts 1828 and

1844, the Game Act 1831 and the Poaching Prevention Act 1862 with a little grafting onto them in modern times (Game Laws (Amendment) Act 1960, Deer Act 1991 and the Firearms Acts 1968–1988). An attempt has been made to unravel the laws by the tables in the Appendix to this chapter setting out the powers of the police and others against poachers under the various statutes.

12.4.2 Theft

Although sporting rights are property, the owner of the land or sporting rights does not own creatures in the wild until they are killed or otherwise brought under his control, even if he reared them. Ownerless things cannot be stolen. Hence the laws against poaching are not about theft but about criminal trespass and possessing the booty or the wherewithal for poaching. However, pheasants in a rearing pen are owned and can be stolen. The Theft Act 1968 summarises the position thus: 'wild creatures, tamed or untamed, shall be regarded as property, but a person cannot steal a wild creature, unless ... it has been reduced into possession by or on behalf of another person' (section 4(4)).

12.4.3 Night poaching

Poaching at night is a more serious offence, carrying heavier penalties than day-time poaching. Under the Night Poaching Acts 1828 and 1844 it is an offence unlawfully by night to take or destroy game or rabbits on open or enclosed lands, highways or gateways. It is a separate offence to enter or to be present on open or enclosed lands (though not a highway) with a gun, net or instrument for taking or destroying game.

'Night' means from one hour after sunset to one hour before sunrise. 'Game' in these Acts means hares, pheasants, partridges, grouse, heath or moorgame, blackgame and bustards (now extinct). Although the second mentioned offence of unlawful entry or presence on land cannot be committed by being on a highway after game, a person with a gun on a highway helping others poaching on private land (such as by keeping a lookout) is equally guilty of the offence (*R v. Whittaker* (1848) 17 LJMC 127).

Three or more armed persons entering land at night for taking game or rabbits commit a more serious offence under section 9 of the 1828 Act, with heavier penalties.

12.4.4 Day-time poaching

It is an offence under section 30 of the 1831 Act to trespass in pursuit of game (as defined for night poaching above), woodcock, snipe and rabbits in the day-time. An extra fine may be imposed if violence is used. The offence is committed whether or not the trespassers find or take any quarry. Any defence to an action for the trespass will be a defence to a charge under section 30.

Sporting Matters

A farm tenant not holding the sporting rights cannot, of course, trespass on the land he occupies, but he commits an offence under section 12 of the 1831 Act, if he unlawfully takes or kills game (*Spicer* v. *Bernard* (1859) 23 JP 311), though not if the booty is woodcock, snipe or ground game.

12.4.5 Armed trespass
Although in some cases it may be doubtful whether the court will be convinced that a plausible defendant was in pursuit of game when trespassing, it is often overlooked that it is an offence under section 20(2) of the Firearms Act 1968 to trespass on land with a firearm. No poaching motive need be proved for a conviction of armed trespass. The Act also makes it an offence to carry a loaded gun in a public place (section 19). A highway, including a public footpath or bridleway, is a public place.

12.4.6 Powers of arrest, seizure, etc.
The steps the police and private persons may take against poachers are summarised in Appendix B. It will be observed that in some instances owners or occupiers of the land and their gamekeepers have powers of arrest, and in some circumstances they may seize from poachers their equipment (but not guns), dogs and any game taken. They (and anyone else having the sporting rights) may require a day poacher to give his name and address and they may require him to leave the private land. An offender who refuses may be arrested.

The power of arrest should be exercised with discretion by the private citizen, possibly with less circumspection by the police. The powers of arrest by constables have been drawn together and clarified in the Police and Criminal Evidence Act 1984, Part III.

12.5 Firearms

12.5.1 Firearms Act 1968
Offences under the Firearms Act 1968 which may be committed by poachers, and the steps that can be taken against offenders under the Act are included in the tables in Appendix B.

12.5.2 Firearms and shotgun certificates
Following the notorious massacre at Hungerford by a mad gunman, the government tightened the laws concerning the possession, keeping and dealing in firearms and shotguns by passing the Firearms (Amendment) Act 1988, which amends the Firearms Act 1968 in a number of respects. Further restrictions were imposed following a major incident in Dunblane which introduced a ban on most handguns under the Firearms Act 1997.

A shotgun licence, but not a firearm certificate, is needed to possess a shotgun lawfully. A shotgun means a smooth-bore gun (not being an air weapon or a revolver gun) with a barrel not less than 24 inches in length, no barrel exceeding 2 inches in diameter and no magazine (or having a non-detachable magazine incapable of holding more than two cartridges)(section 2 1988 Act). Shotgun certificates are issued by the police. They must issue or renew a shotgun certificate if they are satisfied that the applicant can be permitted to posses a shotgun without danger to the public safety or to the peace. The 1988 Act goes on to confuse this injunction by adding that the certificate must be withheld 'if the Chief Officer of police is satisfied that the applicant does not have a good reason for possessing, purchasing or acquiring' a shotgun. It is a good reason to have one if the gun is intended to be used for sporting purposes, competitions or shooting vermin (section 3 1988 Act). Guidance (available from The Stationery Office though currently under review) to the police has been issued by the Home Office dealing with the issue of licences for firearms and shotguns. The police will be concerned to see whether applicants keep their shotguns securely.

A self-loading or pump action shot gun less than 40 inches in length, or with a barrel less than twenty-four inches is a prohibited weapon (section 1 (2) 1988 Act).

It is an offence for anyone under 15 years of age to have with him an assembled shotgun unless while under the supervision of an adult over 21 (unless the gun is covered so that it cannot be fired section 22 (3) 1968). A person under 17 is not allowed to buy or hire a shotgun or its ammunition.

It is an offence for anyone to have in his possession, or to purchase or acquire a firearm (other than a shotgun or air weapon), or ammunition for one, without holding a firearms certificate in force at the time (section 1 1968 Act). Air pistols of more than 6 ft/lbs pressure and air rifles of more than 12 ft/lbs pressure may only be held with a firearms certificate. The police must issue a firearms certificate to an applicant for one, provided they are satisfied he has a good reason for possessing the firearm, he will not be a danger to public safety or to the peace and he is not disqualified from having a firearm (section 27). An army officer serving in Germany was properly refused a firearms certificate, where he owned a house in Warwickshire but let it and did not reside in it (*Burditt* v. *Joslin* [1981] 3 All ER 203).

12.5.3 Firearms and deer

The Deer Act 1991 makes it an offence to use for the purpose of killing, injuring or taking deer any smooth bore gun; or any rifle of less than 0.240 inches calibre or muzzle energy less than 1700 ftlbs; or any air gun, air rifle or air pistol; or any cartridge for a smooth bore gun; or any arrow, spear or similar missile; or any rifle bullet other than soft-nosed or hollow-nosed; or any missile containing poison, stupefying drug or muscle-relaxing agent. It is also an offence to possess any firearm or ammunition for the purpose of committing an offence against the Deer Act 1991 or to possess any trap, snare, poisoned or stupefying bait, net, arrow, spear or prohibited missile for this purpose.

Sporting Matters

12.5.4 Lead shot ban
The Environmental Protection (Restriction on Use of Lead Shot), (England) Regulations 1999 (SI 1999/2170) make the use of lead shot by any person using a smooth bore shot gun of greater bore than 9 mm illegal in the following circumstances

- on or over the foreshore;
- on or over any Site of Special Scientific Interest included in Schedule 1 to the Regulations;

for the purpose of shooting coot, all species of ducks, geese and swans, moorhen, golden plover, and common snipe.

12.5.5 Other weapons, snares, etc.
The 1981 Act has elaborate provisions prohibiting the use of certain weapons and devices for killing, taking, stunning, frightening or decoying birds (section 5) and animals (section 11).

12.6 Fishing

12.6.1 Protecting fishing rights
Fishing rights and fisheries being property, the owner can take legal action against any person, body or authority unlawfully interfering with them. An unauthorised interference with them would give rise to an action in trespass and/or nuisance. Over the centuries the courts have been strong in affording protection. They will give a remedy (usually damages and an injunction) against such people as polluters (for example *Pride of Derby Angling Association* v. *British Celanese* [1953] 1 All ER 179 where the well named pub angling club got injunctions against British Celanese, Derby Corporation and the British Electricity Authority), unauthorised canoeists (*Rawson* v. *Peters* (1973) EGD 259) or careless land drainage authorities (*Welsh National Water Development Authority* v. *Burgess* (1974) RVR 395).

12.6.2 Ownership of fishing rights
Tracing the ownership of fishing rights can involve difficult and expensive investigations through the labyrinths of conveyancing history. Fortunately this is rarely necessary because the law provides aids by way of certain presumptions which prevail unless the contrary is proved. These presumptions are as follows.

(1) *The owner of the river bed owns the fishing.* The law presumes that the owner of the soil of a non-tidal river or a lake owns the fishing rights. Likewise the owner of the fishing is presumed to own the soil

(*Hanbury* v. *Jenkins* [1901] 2 Ch 401). Ownership of the soil and fishing can be severed but the burden of proof is on the party seeking to rebut the presumption.

(2) *Riparian ownership extends to the middle line of the river.* Where a non-tidal river bounds land, the law presumes the riparian owner each side owns the soil (and therefore the fishing) up to the middle line of the river. The middle line is taken at the mean average flow of the river (*Hidson* v. *Ashby* [1896] 2 Ch 1). Again the presumption can be rebutted by evidence to the contrary, but the burden of proof is on the party seeking to displace the presumption. Where both banks are in the same ownership the owner is presumed to have the fishing rights across the river. Where the middle line is the boundary however, whilst in Scotland it is permissible to cast across it (*Fotheringham* v. *Kerr* (1984) 48 P & CR 173), in *Lovett* v. *Fairclough* (1990) 61 P & CR 385, a claim by an owner on the Scottish side of the River Tweed to fish across the middle line in the part of the river owned by the riparian owner on the English side was rejected. Therefore it is not permissible to cast across the mid-point of the river in English law where ownership is split down the middle line of the river.

(3) *The Crown is the owner of tidal rivers.* The law presumes that the Crown owns the soil (and therefore the fishing) of tidal rivers and of the foreshore. This can only be rebutted if the soil or the fishery was disposed of by the King before 1215 when Magna Carta stopped the sovereign doing such things. Riparian ownership of land adjoining a *tidal* river or estuary does not therefore raise a presumption of ownership to the middle line. A river affected by tides is tidal up to the point where ordinary sea tides cease to cause the flow of water to fluctuate horizontally along the banks and vertically up and down them (*West Riding of Yorkshire River Board* v. *Tadcaster UDC* (1907) 97 LT 436).

(4) *Long use and the Prescription Act 1832.* The legal presumption that if a right is exercised for many years without challenge it is lawful, provided it could have had a legal origin, applies to fishing rights. It is known as a *prescriptive right*. Without going into all the complexities of the law of prescription, in outline it is this. The common law presumes the right is lawful and absolute if it has been exercised 'since time immemorial'. It also presumes the right has been enjoyed since time immemorial if there has been 20 years uninterrupted and unchallenged use, but this presumption could be rebutted by proof that at any time after the year 1189 (Richard I's accession) the right did not exist. Also, a claim to a prescriptive right can always be defeated by proving the right was enjoyed under a grant which had

been, or could be, ended (for example, a grant for a fixed period). Under the Prescription Act 1832 the law was modified as follows:

(a) When fishing rights have been enjoyed continuously for 30 years, the claim to them can no longer be defeated by proving there was a time since 1189 when they did not exist, but it could be defeated by disproving the rights in other ways.

(b) When fishing rights have been enjoyed continuously for 60 years, the claim can be defeated only by production of a written grant showing the right has ended or could be revoked.

Neither the public nor any other 'fluctuating and uncertain body' can obtain fishing rights by prescription (*Goodman* v. *Saltash Corporation* (1882) 7 App Cas 633).

12.6.3 Fishing leases

Fishing may be leased with or without land. If it is leased separately from the land it should be by formal deed and the same applies if there is a sale or grant of fishing rights (*Neil* v. *Duke of Devonshire* (1882) 8 App Cas 135). As mentioned above, when there is a lease of land with no reservation of the fishing rights, they are deemed to be included in the lease unless there is clear evidence that the parties intended the contrary (*Browne* v. *Marquis of Sligo* (1859) 10 Ch 1).

It is open to the landlord to include stipulations in the letting regulating the way the fishery is to be enjoyed, such as restricting the numbers of rods, imposing bag and size limits and prohibiting assignment or subletting. If the fishing is let separately from the land (for example to an angling club) it is prudent for the parties to agree and include in the lease the means of access, fishing paths and where the anglers may park vehicles.

12.6.4 Public fishing

The public has a right to fish in tidal waters and on the foreshore, except where the right was lost before Magna Carta 1215 by the King granting the rights away (probably with the land). For the extent of tidal water in rivers see point (3) in section 12.6.2 above. The foreshore is the land between the high and low water marks of ordinary tides, discounting spring and neap tides (*A-Gen* v. *Chambers* (1854) 23 LJ Ch 662).

On the other hand the public has no right to fish non-tidal waters and cannot obtain it (*Smith* v. *Andrews* [1891] 2 Ch 678). On some major rivers public fishing is tolerated in non-tidal reaches and in some places the public have fished for centuries, but even so the acquiescence of the riparian owners does not turn the concession into a right.

The right of public fishing in tidal waters does not automatically carry a right to access across adjoining land. Permission is required to get to the water if there is no public right of way, or else access must be by boat.

12.6.5 Close seasons

The Salmon and Freshwater Fisheries Act 1975 (section 19 and Schedule 1) lays down detailed laws specifying standard close seasons (and weekly close times for salmon and trout) for taking freshwater fish, trout and salmon, with differing close seasons for rod and line fishing, putts and putchers (baskets) and other means (mostly nets). All this is not set out here because in reality the fishery owner and the angler will need to consult the fishery byelaws to ascertain the close season in any place.

The Environment Agency has a duty to make close season byelaws. The byelaws can depart from the standard close seasons, provided they comply with the minimum periods required by the 1975 Act (para 3 Schedule 1). The close seasons often vary from place to place.

It is an offence to fish for, take, kill or attempt to take or kill a fish during the close season for using the method employed for taking the species of fish concerned. However, there are periods when, say, putts and putchers are forbidden for taking salmon, but rod and line fishing is allowed. It is no defence for the landowner or occupier to plead he was fishing on enclosed waters on his own land (unless it is an excepted case, see below).

There are minimum close seasons for different kinds of fish. For salmon the minimum close season for rod and line fishing is 92 days and the weekly close time must be at least 42 hours. For trout the minimum close season for rod and line fishing for trout other than rainbows is 153 days and the minimum close time (only for rod and line) is 42 hours. The rule for rainbows is odd. There need be no close season, but if one is made by byelaw it must be at least 93 days.

For freshwater fish, defined in the 1975 Act as 'any fish living in fresh water exclusive of salmon and trout and of any kinds of fish which migrate to and from tidal waters and of eels' (that is, coarse fish; section 41) the close season may be dispensed with, but if there is one it must be at least 93 days. Nearly always it is the standard close season between 14 March and 16 June (you can fish on both dates mentioned, but not in-between).

With the permission of the Environment Agency and the Department for Environment, Food and Rural Affairs, salmon and trout may be taken during the close seasons for artificial propagation or for some scientific purpose, or, in the case of trout, for stocking waters.

Freshwater fish and rainbow trout have special exceptions

(1) They may be removed by the owner or occupier during their close seasons from any exclusive fishery where salmon or trout are specially preserved
(2) They may be fished for (by rod and line only) with the previous written permission of the owner or occupier, in any such fishery
(3) They may be taken for scientific purposes (no need for official permission)
(4) They may be taken for bait in any exclusive fishery with the written

permission of the owner or occupier, or in any other fishery so long as no byelaw is contravened (sections 19 (3) – (5) and (8) 1975 Act)

There are special rules for eels (1975 Act: sections 2 and 19 (6) (7) and (8) and Schedule 1) and for the removal of fixed devices for taking salmon and migratory trout during the 'nets' close season (section 20). There are also times when selling salmon and trout is prohibited, and when having possession of them for sale between certain dates is unlawful, unless the fish have been canned, frozen, pickled or otherwise preserved (section 22). Restrictions on exporting salmon and trout are in section 23.

12.6.6 Unclean and immature fish

It is an offence 'knowingly' to take, kill or injure (or attempt to) any salmon, trout or freshwater fish which is unclean or immature, though no offence is committed if the fish is taken accidentally and returned to water with the least possible injury (section 2 1975 Act).

'Unclean' means a fish about to spawn, or which has recently spawned but not yet recovered. 'Immature' means fish below the sizes specified locally by byelaws, or a salmon less than 12 inches from the tip of the snout to the fork of the tail.

12.6.7 Fishing licences

It is an offence to take fish for or take fish without a fishing licence in any water where a licence is needed for the fish concerned and the method used (section 27 1975 Act). A fishing licence is always needed for salmon and trout fishing. There are still some places where a licence is not needed to fish for coarse fish.

Fishing licences are issued by the EA. They fix the licence duties for waters in their areas. The Secretary of State's approval is required if there is a written objection to a proposed duty. The duty differs from area to area and it usually varies for trout, salmon and freshwater fish licences. The licence will state what kind of fishing it authorises and for what kind of fish (separate licences are needed for rod and line, basket and net fishing) and the period of validity, which may be, for example, a day, week, month or year (see Part IV of and Schedule 2 to the 1975 Act for fishing licences).

A rod and line licence also authorises the use with it of a gaff, tailer or landing net (section 25(4) 1975 Act). A salmon fishing licence also authorises fishing for trout, and a salmon or trout licence also authorises fishing for freshwater fish and eels (section 25 (5) and (6)). The procedure the Environment Agency follows for fixing or altering licence duties is laid down in schedule 2 to the 1975 Act. The points to note with relevance to fishing licences are as follows.

(1) A fishing licence does not confer a right to fish. The angler must also have the fishing rights or the owner's permission to fish (paragraph 16 Schedule 2 1975 Act).

(2) A rod and line licence is not transferable (section 25(2)), and it cannot take effect before the moment it is issued (*Wharton v. Taylor* (1965) 109 SJ 475 (DC)).
(3) Anyone tendering the duty is entitled to a licence unless disqualified by a court for fishing offences, or unless an order has been made limiting the number to be issued (section 26 and paragraph 15 Schedule 2).
(4) Any person holding (and producing) a fishing licence, or any constable or official bailiff, may require anybody found fishing to produce his licence and to give his name and address. Failure to comply is an offence (section 35).

A person or association entitled to an exclusive right of fishing may be granted a general licence for the fishery under which the licence holder or any person authorised in writing by him (or by the secretary of the association) may fish there, subject to any conditions agreed with the Environment Agency (section 25(7)). The duty payable is negotiated with the Agency. A general licence is useful where the landowner wishes to allow guests to fish who may not hold fishing licences. The Environment Agency may not withhold a general licence without a good reason (*Mills v. Avon and Dorset River Board* [1953] 1 All ER 382).

12.6.8 Illegal fishing methods

It is an offence to use any of the 'instruments' or modes of fishing prohibited by the 1975 Act. The prohibited instruments are any firearm; an otter (a board used for carrying bait), lath or jack, wire or snare; a crossline or setline; a spear, gaff, stroke-haul, snatch or other like instrument; and a light. With the exception of gaffs and tailers, none of these instruments may be used for taking or killing salmon, trout or freshwater fish, nor may they be in the possession of anyone with the intention of so using them (section 1). A gaff or tailer may be used as auxiliary to angling with rod and line, or be possessed for that purpose (section 1(4)).

Definitions of the prohibited instruments are given in the 1975 Act. The instruments will be familiar to poachers. They are mainly devices for foul-hooking fish, or for running out baits or lures, or to be left across or in waters without rods.

The Act prohibits the shooting or working of any seine or draft net for salmon or migratory trout across more than three-quarters of the width of any water and there are restrictions on the size of mesh (section 3).

It is an offence to throw or discharge any stone or other missile for the purpose of taking or killing, or facilitating the taking or killing of any salmon or trout or freshwater fish (section 1(1)). It is also illegal to use any fish roe for bait (section 2); or to use any explosive or poisonous substance or any electrical device with the intent to take or destroy fish, unless it is done with the written permission of the Environment Agency for scientific purposes, or for protecting, improving or replacing fish stocks (section 5).

12.7 Poaching of fish

12.7.1 Statutory offences

Just as there are special statutory offences for poaching game because of difficulties about proving theft by poachers, as explained above (see section 12.4 above) the same applies to illegal fishing. Offences are laid down by the Theft Act 1968, which deals separately with night and daytime poaching.

12.7.2 Night poaching

It is an offence unlawfully to take or destroy at night fish in a private fishery, or water in which there is a private right of fishing. It is also an offence to attempt to. 'Night' is from one hour after sunset to one hour before sunrise. To some extent this is not only a night offence, because the Act states that the offence is committed if it is done (or attempted) in the day-time by a means other than angling (section 32(1) Theft Act 1968 and paragraph 2 Schedule 1).

12.7.3 Day-time poaching

If the unlawful taking or destruction of fish, or an attempt at same, is done by angling in the day-time, a different offence is committed for which the penalties are lighter (section 32(1) and paragraph 2 Schedule 1 Theft Act 1968).

12.7.4 Remedies against poachers

The old remedy of confiscating the poacher's tackle is no longer allowed, but 'any person' may seize for production in court anything the poacher has with him for taking or destroying fish, and the court may order forfeiture of it. Any person may also arrest without a warrant anyone caught committing the offence described above under 'night poaching' (including the non-angling offence by day), but not offenders poaching by angling in the day-time. The powers of fishery bailiffs are unaltered by the Police and Criminal Evidence Act 1984.

The poacher will also be a trespasser and liable in the civil courts to an action for damages and an injunction.

The Salmon Act 1986 has created a new offence of 'handling' illegally taken salmon in suspicious circumstances, and provides for a scheme of dealer licensing to be brought in by order but none is being made.

12.8 The Environment Agency

The responsibilities placed on the sometime water authorities by the 1975 Act were transferred to the National Rivers Authority by the Water Act 1989, and then to the Environment Agency from 1 April 1996 by the Environment Act 1995 (see Chapter 11). The EA's statutory duties include a duty 'to maintain, improve and develop salmon fisheries, trout fisheries, freshwater fisheries and eel fisheries' in consultation with statutory fishery advisory committees (sections 6(6) and 13 Environment Act 1995).

12.9 Hunting

12.9.1 Trespass

It was once widely believed a hunt could cross land without permission. It cannot lawfully do so. Members of a hunt entering land without permission commit a trespass (*Paul* v. *Summerhayes* [1874] 4 QBD 9 (DC)) and a Master of Hounds, or other hunt official, may be held liable for trespass even if he does not enter the land himself (*Robinson* v. *Vaughan* (1838) 8 C & P 252) though not if entry was made against his will (*Baker* v. *Berkeley* (1827) 3 C & P 32). Although a Master of Hounds is only liable for trespass if he intends hounds to enter land unauthorised, or if he negligently fails to prevent it, persistent hunting close to land where it is effectively impossible to prevent hounds entering can be evidence of an intention to trespass (*League Against Cruel Sports* v. *Scott* [1985] 2 All ER 489). A member of a hunt holding no official position cannot be held responsible for the trespass of the other hunters (*Paget* v. *Birkbeck* (1863) 3 F & F 683).

Trespass is also committed by allowing or sending hounds on to another's land without permission, even if no person enters the land (see *Read* v. *Edwards* (1864) 17 CBNS 245).

It is necessary, therefore, for a hunt to get prior consent from the occupiers before they can hunt over any land. This, of course, is the usual practice and arrangements are made about compensation for damage done.

The pursuit of hunting has become increasingly controversial recently, the Government has promised a free vote on the issue in accordance with the labour party's General Election Manifesto. However, at the time of writing (2002) it is unclear what form the Bill will take, and when/whether the Government will provide parliamentary time for the issue to be settled. There has also been increased agitation between participators and protesters, in relation to which section 68 of the Criminal Justice and Public Order Act 1997 (aggravated trespass) may be invoked – see section 10.8, Chapter 10.

12.9.2 Fresh pursuit

The Game Act 1831, section 35, provides that the offences of trespassing in pursuit of game do not apply to anyone hunting with hounds (or coursing with greyhounds) in fresh pursuit of any deer, hare or fox already started upon some other land. Although the section gives a defence to a charge of poaching, it is no defence to an action of trespass (*Paul* v. *Summerhayes*, above).

Table 1 Arrest

Powers to apprehend poachers: search, seizure, confiscation, etc.

Firearms Act 1968 Police and Criminal Evidence Act 1984	Night Poaching Acts 1828 and 1844 Game Laws (amendment) Act 1960 Police and Criminal Evidence Act 1984	Offences under Game Act 1831 Wild Creatures and Forest Laws Act 1971 Police and Criminal Evidence Act 1984	Deer Act 1991

Trespass on land with a firearm (1968) Act s. 20).
 	
Carrying a firearm in public place (including highway) (s. 19).
 	
Having small calibre pistol outside premises of licensed pistol club (s. 19A).
 	
Failure to produce certificate (s. 48).
 	
A constable has powers of arrest under s. 25, Part III of the Police and Criminal Evidence Act 1984. A constable has powers of entry and search under s. 46, Firearms Act 1968. He has power to demand production of firearm certificate of shotgun certificate (s. 48, 1968 Act). If it is not produced, a constable may ask for the person's name and address.
  Failure to provide the information or providing false information is a ground for arrest.
	Police constables have powers of arrest under the Police and Criminal Evidence Act 1984.
	
	Owners, occupiers, lords of the manor, their gamekeepers and servants may apprehend offenders found on any land committing an offence under s. 1 of the 1828 Act. If pursuit is made, the offender can be apprehended in the place to which he has escaped (s. 2, 1828 Act). (S. 1 makes it an offence to take or destroy game or rabbits by night or enter land for taking game.) If an offender assaults a person trying to arrest him, he is liable to 6 months imprisonment, or a fine, or both (s. 2, 1828 Act).
	  The Act also applies to taking rabbits or game on the highway – and adjoining owners, occupiers, etc. can arrest as above (1844 Act).
	
	The Game Laws (Amendment) Act 1960 gives police constables powers of entry for purposes of arrest under the Police and Criminal Evidence Act 1984.
		Any person found trespassing on land in search of game in the day-time may be required to leave and give his name and address by any of the following
		(a) the person who enjoys the sporting rights
		(b) the occupier
		(c) any gamekeeper or servant of either of the above
		(d) any police constable
		
		Any offender who refuses to give his name and address may be arrested by any of the above (1831 Act ss. 31, 31A).
		
		A police constable has powers of entry under the Game Laws (Amendment) Act 1960, s. 2, for the purpose of ss. 31, 31A above and to arrest offenders under s. 25, Police and Criminal Evidence Act 1984.
			A constable may arrest without warrant a person suspected of committing an offence under the 1991 Act (s. 12) and for the purpose may enter any land other than a dwelling house.
			
			A constable has powers of arrest where he suspects an offence is being committed, under s. 25, Police and Criminal Evidence Act 1984.

Table 2 Seizure of guns, etc.

Firearms Act 1968	Night Poaching Act 1828 Game Laws (Amendment) Act 1960 Police and Criminal Evidence Act 1984	Game Act 1831	Deer Act 1991	Poaching Prevention Act 1862 [as amended by Game Laws (Amendment) Act 1960]
Power of constables to stop and search (s. 47). A constable who suspects that a person is committing armed trespass on private land under s. 20 may require him to hand over the firearm or any ammunition for examination. A constable may also search and detain a person he suspects is committing or about to commit armed trespass on private land under s. 20. The same powers may be exercised with respect to a person suspected of having a firearm in a public place under s. 19. A person convicted of an offence under ss. 19, 20, 48 may have his firearm forfeited by the court and his firearm certificate may also be cancelled (s. 52). s. 47(4). A constable has powers to stop and search vehicles which he suspects are being used for armed trespass under s. 20.	Where a person is apprehended in accordance with s. 25 of the Police and Criminal Evidence Act 1984 for an offence under ss. 1, 9 of the Night Poaching Act 1828, a police constable by or in whose presence he was apprehended may search, seize and detain any (a) game or rabbits (b) gun, part of gun, or cartridges or other ammunition (c) nets, traps, snares, etc. (s. 4, 1960 Act, as amended by Schedule 6 of 1984 Act) If a person is convicted, the court may order the forfeiture of any or all of the above items.	Lords of the manor may appoint gamekeepers to act within limits of the manor (s. 13). Such gamekeepers may be authorised to seize and take dogs, nets, and other engines and instruments for the killing or taking of game within the manor by persons not having a game licence. This does not give power to seize guns but includes snares. What is seized under this section may be kept. Game may be demanded from trespassers and seized if not delivered up when demanded (s. 36). This right is for the person enjoying the sporting rights, the occupier of the land, and their gamekeepers and servants. Where a person is apprehended under s.31 (see note above), a police constable by or in whose presence he was apprehended may search him and seize any	Where a person is convicted of an offence under the Act, the court may order the forfeiture of any vehicle, animal or other thing used or capable of being used to take, kill or injure the deer (s. 13 1991 Act), and any deer which was the subject of the offence. A constable may seize any vehicle, animal, weapon, etc. under s. 12 1991 Act.	Constable has power in any highway, street or public place to search any person he suspects has been poaching and has in his possession any poached game, gun, or part of a gun (s. 2). A constable also has power to stop and search any vehicle which he suspects is carrying game, and if he finds game, a gun or part of a gun, he can seize and detain them. By the Game Laws (Amendment) Act 1960, the section is made to apply to cartridges and other ammunition and to nets, traps, snares, etc. Where a person is convicted under this section, a court may order forfeiture of game, guns, etc. seized under the section (s. 3, 1960 Act).

Cont.

Table 2 *Continued*

Firearms Act 1968	Night Poaching Act 1828 Game Laws (Amendment) Act 1960 Police and Criminal Evidence Act 1984	Game Act 1831	Deer Act 1991	Poaching Prevention Act 1862 [as amended by Game Laws (Amendment) Act 1960]
s. 48(2). A constable may seize and detain firearm, ammunition and shotgun if a person refuses to produce his certificate. Powers of search with warrant. A constable with a warrant granted under s. 46 may enter premises, if necessary by force, and search the premises or place and every person found there. He may also seize and detain any firearm or ammunition.		(a) game or rabbits (b) gun or part of a gun, or cartridges, or other ammunition (c) nets, traps, snares, etc. If the person is convicted, the court *may* order the forfeiture of any or all of the above items (s. 4 of 1960 Act).		

Table 3 Fines and other penalties

Firearms Act 1968	Night Poaching Acts 1828 and 1844 (amended by Game Laws (Amendment) Act 1960 and Criminal Law Act 1977)	Game Act 1831 (as amended by Wild Creatures and Forest Laws Act 1971 and Game Act 1970)	Poaching Prevention Act 1862 [as amended]	Deer Act 1991
Schedule 6. Part I Possessing etc. firearm or ammunition without firearms certificate (s. 1). Summary conviction – six months or a fine or both. On indictment – aggravated offence 5 years or fine or both in other cases 3 years or a fine or both. Possessing shotgun without certificate, (s. 2(1)). On summary conviction 6 months or a fine or both. On indictment – 5 years or a fine or both. Use of firearm to resist or prevent lawful arrest. On indictment: life imprisonment (s. 17) or fine or both.	Taking or destroying game or rabbits by night or entering land for that purpose (s. 1 of 1828 Act). Penalty: a fine. Assaults by persons committing offences under the Act (s. 2). On summary conviction – 6 months or a fine or both (offence triable only summarily). Entering land with others armed and for the purpose of taking or destroying game or rabbits (s. 9). On summary conviction – 6 months or a fine or both. The above penalties applying to persons destroying rabbits or game at night, apply to persons destroying rabbits or game by night on any public road, highway, path and openings and gates leading onto them (s. 1 Night Poaching Act 1844).	Killing or taking game on Sunday or Christmas Day (s. 3). Fine plus costs. Taking or killing game in the close season. Fine plus costs. For laying poison to kill game. Fine plus costs. For killing or taking game without a licence. Fine plus costs (see also Game Licence Act 1860, s. 4). Trespassing in the day-time in search of game woodcock, snipe or rabbits. Fine plus costs. Trespass as above by five or more persons. Fine plus costs (s. 30). Trespass as above by five or more persons using violence. Fine in addition to any other penalty (s. 32).	Where a person searched by constable on highway, street or public place under s. 2 of Act is found to have game unlawfully obtained on land, or gun or part of gun for unlawfully obtaining game. Penalty: a fine.	Any offence under the Act. The penalty is a fine or 3 months imprisonment or both.

Cont.

Table 3 *Continued*

Firearms Act 1968	Night Poaching Acts 1828 and 1844 (amended by Game Laws (Amendment) Act 1960 and Criminal Law Act 1977)	Game Act 1831 (as amended by Wild Creatures and Forest Laws Act 1971 and Game Act 1970)	Poaching Prevention Act 1862 [as amended]	Deer Act 1991
Carrying loaded firearm in public place – s. 10. On summary conviction – 6 months or a fine or both. Indictment (except for air weapon) 7 years or fine or both.				
Trespassing with a firearm on land (s. 20(2)). On summary conviction – 3 months or a fine or both.				
Failure to hand over firearm or ammunition on demand by a constable (s. 47(2)). On summary conviction – 3 months or a fine or both.				

Chapter 13
Environment

Abbreviations in this chapter
'1990 Act'   Environmental Protection Act 1990
'1991 Act'   Water Resources Act 1991
'Secretary of State'   Secretary of State for Environment, Food and Rural Affairs, or the National Assembly for Wales where the context requires

13.1 Introduction

The protection of the environment and pollution occasioned by industry and other modern activities have been given increased prominence in UK legislation in recent years. Given the highly polluting nature of agricultural waste and processes, this increased prominence is very evident in the agricultural sector. Agricultural pollution is dealt with by legislation in a number of ways, both through criminal sanctions to deter incidents, and with a range of measures to deal with more diffuse problem situations (these include contractual agreements with farmers and precautionary regulations). The wealth of law in this area is substantial, and increasing. Below are highlighted some of the main aspects likely to be of relevance in the countryside.

13.2 Water pollution

13.2.1 Water pollution offences

Section 85 of the 1991 Act makes it a criminal offence to either 'cause' or 'knowingly permit' any poisonous, noxious or polluting matter, or any solid waste matter, to enter controlled waters. It is also an offence to cause or knowingly permit any trade or sewage effluent to be discharged, in breach of a prohibition notice, onto any land or into any waters of a lake or pond. A breach of section 85 carries a fine of up to £20 000 and/or up to two years imprisonment.

The terms of this offence have been construed strictly by the courts. In *Alphacell Ltd* v. *Woodward* [1972] 2 All ER 475 HL, it was established that 'causing' and 'knowingly permitting' are two distinct and separate grounds of liability. Knowledge or intention are irrelevant to the offence of 'causing' water pollution – all that is required for the offence being a factual causal link between the agricultural activity and the pollution. The chain of causation will only be broken if there is an intervening event, such as an Act of God or third party intervention, such that the action of the defendant was not the causative factor in the pollution incident, but rather part of the background circumstances. 'Knowingly permitting' water pollution means failure to act to prevent pollution, in circumstances where it is within the power of the accused to take affirmative steps to avoid the incident complained of, together with knowledge that pollution will result from the failures to act. This was considered in *Empress Car Co. Ltd* v. *NRA* [1977] Env. LR 227, where the release of a diesel tap by an unknown party was still held to be the responsibility of the occupier *inter alia* because the tap was not locked and was accessible to the public and therefore the procedures for controlling any overflow were quite inadequate.

The Environment Agency is given power under the 1991 Act to issue discharge consent and prohibition notices controlling effluent discharges in certain situations: see sections 86 and 88 and Schedule 10. It is then not an offence under section 85 in respect of the entry of any matter into any waters if made in accordance with a discharge consent, or indeed specified other consents such as waste management or disposal licences (see below).

13.2.2 Preventative measures

Diffuse water pollution is a more difficult problem to regulate, and the recent emphasis has been on preventative measures, i.e. requiring certain standards for some farm practices, with criminal sanctions for breaches of the same, thus reducing the likelihood of escape of pollutants into watercourses. Under what is now section 92 of the 1991 Act, the Secretary of State is empowered to make regulations prohibiting a person from having control of polluting or noxious matters unless prescribed steps have been taken to prevent pollution, under which the Control of Pollution (Silage, Slurry and Fuel Oil) Regulations 1991 (SI 1991/324 as amended) were introduced. These set down standards for the storage of silage in silos, bales or clamps, for the storage of slurry, and for the storage of fuel oil. However, the regulations are not retrospective, and storage facilities in use prior to these regulations are excluded.

Essentially the regulations provide that silage can only be made in a silo with an impermeable base surrounded by a system to collect any effluent and take it to a storage tank. Silage can also be made in bales and a tower silo of required standards. Any other methods of silage-making must be registered with the Environment Agency and may continue for five years, unless the Environment Agency considers there is a serious pollution risk.

Farms that produce slurry must have a minimum four months storage capacity unless the Environment Agency is satisfied that less capacity does not represent a serious pollution risk. There are also minimum standards for loading and minimum heights for slurry lagoons.

Tanks for storing more than 1500 l of fuel on a farm must be surrounded by an impervious wall called a bund to prevent leaks reaching watercourses.

13.2.3 Nitrates sensitive areas and water protection zones

Section 93 of the 1991 Act provides powers for areas to be designated water protection areas, thus enabling the Environment Agency to have control over the application of pesticides and other potential pollutants in the surrounding areas. Sections 94 and 95 provide for the establishment of the *nitrate sensitive areas* (NSAs) for England and Wales in which controls can be introduced over agricultural activities to reduce the amount of nitrate leaking from agricultural land into water sources. Zones in north east and central England have been formally designated NSAs. The NSA scheme involves farmers making changes going well beyond good agricultural practice, and accordingly compensation is payable in return for a management contract regulating the farmers' land management practices.

There are also provisions for designation of nitrate vulnerable zones. These are to be found in the Action Programme for Nitrate Vulnerable Zones (England and Wales) Regulations 1998 (SI 1998/1202). These are based on a regulatory model, breach of which is a criminal offence. The standards required by the regulations are similar to 'good agricultural practice', and therefore no compensation is payable.

Together, these two programmes were intended to implement the European Union Nitrate Directive 91/676/EEC. However, following a reference to the European Court of Justice, the UK was found not to have fully implemented this Directive. At the time of writing DEFRA is consulting on the further implementation of the 1991 Directive.

Section 93 of the 1991 Act also contains a more general power for the Secretary of State to designate a land a 'water protection zone'. This may be done if it is necessary to prevent or control the entry of any poisonous noxious or polluting matter (other than nitrates) into controlled waters, and to prohibit or restrict the carrying of activities likely to lead to pollution within the area designated. To date, these have only been exercised in respect of the River Dee catchment area (see SI 1999/916).

13.3 Other pollution control measures

13.3.1 Integrated pollution control

The emission of noxious substances other than smoke is regulated by the system of integrated pollution control to be found in Part I of the 1990 Act. The prescribed processes which require specific authorisation are set out in the Environment Protection (Prescribed Processes and Substances) Regulations 1992 (SI 1991/472 as amended by SI 1992/614). Most prescribed processes likely to apply to agriculture are *Part B cases* and therefore under local authority control. Examples might be the burning of fuel manufactured from or including waste (e.g. poultry litter combustion processes), the incineration of animal remains, and the treatment and/or processing of animal or vegetable matter.

13.3.2 Protection of the air

Various Acts and regulations relating to clean air may apply to farms. For instance the Clean Air Act 1956 requires any farm that is within a smoke controlled area to comply with sections 11–15 of the Act, which prohibit the emission of smoke from any building or chimney and require the use of authorised fuels. Section 1 of the Clean Air Act 1958 prohibits the emission of dark smoke from industrial or trade premises, and this includes agricultural and horticultural land. The Clean Air (Emission of Dark Smoke Exemption) Regulations 1969 permit the burning in certain circumstances of various farm wastes which include the burning of animal or poultry carcases and pesticide containers. Other exemptions include timber and other waste matter which result from the demolition of the building or clearance of a site in connection with building operations. As mentioned above, certain provisions of Part I of the 1990 Act could apply to farms.

13.3.3 Pesticides

There are numerous provisions and regulations relating to the use and effects of pesticides, just a few are mentioned here. Part III of the Food and Environment Protection Act 1985 regulates the use of pesticides that threaten the health of human beings creatures and plants and so as to safeguard the environment and secure the humane control of pests. Following this Act the Advisory Committee on Pesticides was established and various codes of practice have been issued, among others, for suppliers of pesticides to agriculture, for the safe use of pesticides on farms and holdings, for the prevention of spray drift, and for the disposal of waste pesticides and containers. There are also additional controls on the use of pesticides in environmentally sensitive areas designated under the Agricultural Act 1986.

13.3.4 Other provisions

See also the sections on stubble burning and statutory nuisances in Chapter 10 at under section 10.5.

13.4 The codes of practice

13.4.1 Status of codes

There are at present three codes of practice relating to the protection of air, water and soil, obtainable free of charge from DEFRA. These are not directly legally binding, but in some cases may have some indirect legal status, or otherwise may in practice assist with the avoidance, or in defence, of a prosecution.

13.4.2 The Code of Good Agricultural Practice for the Protection of Air (1993)

This gives detailed guidance to producers as to the law, pollution problems arising from many farming activities, and technical advice on how to avoid contravening the law on pollution control. This code has no legal effect, direct or indirect – it is simply advisory.

13.4.3 The Code of Good Agricultural Practice for the Protection of Water (1991, revised 1997)

This code has indirect legal enforceability. By virtue of section 97 of the 1991 Act, a contravention of its terms must be taken into account by the Environment Agency when deciding how and when to exercise its statutory functions, when issuing prohibition notices to prevent discharges of trade effluent, and when exercising its powers under the Control of Pollution [etc.] Regulations 1991. However, breach of the code does not of itself involve the committing of an offence; conversely neither does compliance with it constitute a defence under section 85 of the Act. Therefore the code has some indirect legal effect, and compliance with its terms will render prosecution for a pollution incident less likely.

13.4.4 The Code of Good Agricultural Practice for the Protection of Soil (1994)

This is not a statutory code, but is more a practical guide giving producers advice and information both as to how to avoid causing long-term derogation of the soil, and to maintain soil fertility levels at the optimum level to support plant life. However, there are some legal implications of such activities, for instance under the Town and Country Planning Act 1990. The application of sewage sludge to land as a fertiliser is regulated under the Sludge (Use in Agriculture) Regulations 1989 (SI 1989/1263, as amended). These prevent the release of sewage sludge with concentrations of heavy metals exceeding specified limits and prohibit the grazing of animals on land for three weeks after sewage sludge has been applied. Fruit and other crops normally eaten raw cannot be harvested for ten months after an application of sewage sludge.

13.5 Contaminated land

13.5.1 Introduction of regime

The contaminated land regime was inserted as Part IIA of the Environmental Protection Act 1990 by section 57 of the 1995 Act. The new sections are 78A to 78YC. The regime relies heavily upon regulations and statutory and non-statutory guidance. It came into effect in England on 1 April 2000. The Guidance is given in DETR Circular 2/2000 *Environmental Protection Act 1990: Part IIA Contaminated Land*. Annex 2 of the

Guidance provides a summary of the regime, and statutory guidance is contained in Annex 3.

13.5.2 Administration
The regime is administered by local authorities (unitary councils or district councils) and by the Environment Agency. Primary responsibility for identifying sites and securing their remediation rests with local authorities. The Environment Agency takes responsibility for special sites, those with particular difficulties, and military and nuclear sites. The 'enforcing authority' is the body responsible for regulating a particular site, which could be the local authority or the Environment Agency.

13.5.3 Statutory guidance
The 1990 Act provides for the Secretary of State (for England) and the National Assembly for Wales to issue two forms of guidance

(1) Guidance with which the local authority *must* act in accordance. This covers the definition and identification of contaminated land and exclusion from, and apportionment of, liability for remediation
(2) Guidance to which the local authority must have *regard*. This is the remediation of contaminated land and recovery of the costs of remediation

(1) above imposes a duty on the local authority to follow the guidance. In this respect the guidance has the status of law, and its interpretation could be a matter for the court.

13.5.4 Definition of contaminated land
Section 78A(2) defines 'contaminated land' as

> '... any land which appears to the local authority in whose area it is situated to be in such a condition, by reasons of substances in, on or under the land that –
> (a) significant harm is being caused or there is a significant possibility of such harm being caused; or
> (b) pollution of controlled waters is being, or is likely to be, caused.'

Subsection (5) provides that all the key terms, including the meaning of significant harm, significant possibility etc. are to be determined in accordance with the statutory guidance.

The guidance provides that for any land to be contaminated there must be a pollutant linkage that is

> '❐ a contaminant (a substance with the potential to cause harm/ pollution);

- which is harming or could harm a receptor (a living organism, ecological system or property or controlled waters);
- via a pathway (a route or means)'

Guidance, Annex 3, A.12–15

'Substance' means 'any natural or artificial substance, whether in solid or liquid form or in the form of a gas or a vapour' (section 78A(9)).

'Harm' means 'harm to the health of living organisms or other interferences with the ecological system of which they form part and, in the case of man, includes harm to his property' (section 78A(4)).

The local authority must be satisfied that the pollutant linkage

'- is resulting in significant harm;
- presents a significant possibility of significant harm; or
- is resulting in, or is likely to result in, pollution of controlled waters.'

Guidance Annex 3, para A.19

The guidance provides that the following types of harm are to be regarded as significant (paragraph A.23, Table A).

- Death, disease, serious injury, genetic mutation, birth defects, or the impairment of the reproductive functions of humans (a 'human health effect').
- Irreversible or other substantial adverse change in the functioning of the ecological system within specified locations or which affects any species of special interest within that location and endangers the long-term maintenance of its population there (these include SSSIs, national nature reserves, local nature reserves, marine nature reserves, special protection areas, etc.).
- Death, disease or other physical damage to crops, produce for consumption, livestock, other owned or domesticated animals, or wild animals the subject of shooting or fishing rights such that there is a substantial loss in value (or diminution in yield). A 20% loss or diminution is significant. No loss of value is needed for domestic pets.
- Structural failure, substantial damage or substantial interference with any associated property right in respect of buildings. This occurs when any part of the building ceases to be able to be used for the purpose for which it was intended.

Pollution of controlled waters is 'the entry into controlled waters of any poisonous, noxious or polluting matter or any solid waste matter' (section 78A(9)).

13.5.5 *Local authority inspection and strategy*

Section 78B requires each local authority to inspect their area from time to time for the purpose of identifying contaminated land and to determine which land should be designated as a *special site*, i.e. one so seriously contaminated that it falls within the jurisdiction of the Environment Agency. The guidance instructs local authorities to prioritise the most serious cases first.

Local authorities should prepare a written strategy within 15 months of the issue of the guidance. A determination that land is contaminated should be recorded in writing with a description of the pollutant linkage (including pollutant, pathway and receptor) and summaries of evidence, assessment and compliance with the guidance.

13.5.6 *Service of remediation notices*

Where a site has been designated as a special site or has been identified as contaminated land, the enforcing authority has to serve a remediation notice on each appropriate person 'specifying what that person is to do by way of remediation and the periods within which he is required to do each of the things as specified' (section 78E(1)). This duty is subject to a number of restrictions which mean that the serving of every remediation notice is a last resort, rather than a first. Having identified the contaminated land, the local authority must notify:

- The Environment Agency
- The owner of the land
- Any occupier of the land
- Any person who appears to be an 'appropriate person' (as to which see below)

identifying in each case the capacity in which they are served.

The enforcing authority is obliged to consult these persons and may not serve a remediation notice until three months have elapsed, unless there is imminent danger of serious harm, or serious pollution of controlled waters being caused (section 78H).

No remediation notice is served if

(1) the only things which may be done are not reasonable having regard to the cost involved, the seriousness of the harm or pollution and the guidance on this issue;
(2) the appropriate things are being done or will be done without a remediation notice;
(3) the notice would have to be served on the authority itself, or
(4) the enforcing authority has powers under section 78N to carry out remediation itself.

See section 78H

If no notice is possible by reason of (1) above, the authority must prepare a remediation declaration. If (2) applies, the person who will do the work must prepare a remediation statement. If (3) or (4) apply, the authority must prepare a remediation statement.

Remediation action may consist of assessment action (establishing what remedial treatment is required), remedial treatment action (to reach a standard of remediation) and monitoring action (to ascertain any changes to a pollutant linkage).

An 'appropriate person' is determined in accordance with the criteria in section 78F which provides for two categories. The first category is the *Class A person*, which is the person who caused or knowingly permitted the substances or any of the substances by reason of which the contaminated land in question is such land. If no Class A person can be found, then resort is made to the *Class B person*, which is the present owner or occupier of the contaminated land. This potentially has huge implications for landowners, with the possibility of contamination going back to the last century, and therefore no realistic chance of tracking down the original polluter.

The guidance goes on to give detail on the process of identifying appropriate persons, and the approach where two or more persons are liable. Owners or occupiers (Class B persons) are excluded if they occupy under a licence with no market value or pay a rack rent for the property (para D of the guidance). Liabilities are apportioned under Class A to reflect the relative responsibility of each person. Class B liability is apportioned by capital value (calculated on the assumption of the absence of contamination).

13.5.7 Rights of appeal

Recipients of remediation notices have a right to appeal within 21 days to the magistrates' court where the local authority has served a notice and to the Secretary of State where the Environment Agency has served the notice (section 78L). Whilst there are 19 grounds of appeal in total, these are mostly confined to the enforcing agency exceeding their jurisdiction for not following the guidance, etc.

13.5.8 Penalties

Failure to comply with a remediation notice without reasonable excuse is an offence triable in the magistrates' court. The maximum penalty is a level 5 fine (presently £5000) unless the notice relates to industrial, trade or business premises in which case the maximum fine is £20 000. If the enforcing authority is of the opinion that prosecution would not afford an effectual remedy, it may apply to the High Court for an injunction. The enforcing agency may also carry out remediation and recover the cost from the appropriate person.

13.5.9 Contaminated land register

Enforcing authorities must maintain a contaminated land register detailing its action under the regime. The register requires particulars of remediation notices, remediation declaration, remediation statements and notifications of claim remediation. It should also detail where a site is contaminated land, but has been dealt with under integrated pollution control or waste management licensing powers.

13.5.10 Civil liability
It should not be forgotten that an action in nuisance could be brought by one owner against another owner if their land is affected by the latter's activities (see Chapter 10, section 10.5).

13.5.11 Additional statutory powers
In addition to and parallel with the contaminated land régime the Environment Agency can use work notice procedures under Schedule 22 of the 1990 Act to deal with serious water pollution. Essentially it can take enforcement action to prevent or deal with pollution of water, and charge for the costs of this.

13.5.12 Buying and selling contaminated land
With the contaminated régime having the potential to blight the value of land, it goes without saying that environmental investigations will form a key part of the advice to be provided by professionals acting in connection with a land transaction. Environmental questions will form an increasing part of precontractual inquiries, and local authority searches may provide clues as to environmental risk. If there are grounds for supposing the land may be contaminated then a full environmental audit may have to be carried out. The implications of the regime will also affect the terms of the granting of tenancies, with the risk of a landowner being liable for a tenant's actions on the land long after the tenancy had ended.

13.6 Fly tipping

13.6.1 The offences
Various offences exist in relation to fly tipping. The offences under sections 33 and 34 of the 1990 Act in relation to waste management licences (see section 13.7.3 below) could apply, although if the fly tipping was done by an unknown third party then there would be a defence afforded to the occupier. It should be noted that land contaminated by fly tipping is exempt from the contaminated land regime.

Under section 2 of the Refuse Disposal (Amenity) Act 1978 the deliberate abandonment of any matter on land in the open air or on any other land forming part of a highway is an offence, punishable by fine and/or imprisonment.

Under section 148 of the Highways Act 1980 it is an offence to

- Deposit dung, compost or other material for dressing land or any rubbish either on or within 15 feet from the centre of a made-up carriageway
- Deposit anything on a highway (which includes footpaths bridleways and byways) to the interruption of any user of the highway

Under section 130 of the Highways Act 1980, it is the duty of the highway authority to prevent, as far as possible, the stopping up or obstruction of the highways for which it is responsible and other highways if in its opinion, the stopping up or obstruction of the highway would be prejudicial to the interests of its area. It is also the duty of a highway authority to prevent any unlawful encroachment on any roadside waste comprised in a highway for which it is responsible.

13.6.2 Who deals with fly tipping – the relevant powers and duties

13.6.2.1 Vehicles

Under sections 3–5 of the Refuse Disposal (Amenity) Act 1978 the local authority is under a *duty* to remove motor vehicles unlawfully abandoned on any land in the open and can recover its charges from either

- the owner, unless he can show that he was not concerned in and did not know of its being abandoned; or
- the person who abandoned it.

For these purposes a *motor vehicle* includes a trailer for use as an attachment to a vehicle, and any chassis or body with or without wheels appearing to have formed part of a vehicle or trailer.

13.6.2.2 Articles other than vehicles

Where there has been a breach of section 33 of the 1990 Act (see section 13.7 below) the waste regulation authority (the Environment Agency) or the waste collection authority (usually the local authority) may serve a notice on the occupier requiring him to remove the waste and/or eliminate or reduce the consequences of the deposit (section 59 1990 Act). It is an offence to fail to comply with such a notice. However, the occupier may appeal within 21 days from service of the notice to the magistrates' court on the basis that he neither deposited nor knowingly caused nor knowingly permitted the deposit of the waste, or that there is a material defect in the notice.

Where the occupier fails to adhere to the notice provisions (and has not appealed), the Environment Agency or the local authority may fulfil the notice requirements and recover their reasonable costs from the occupier.

The waste may be removed by the Environment Agency or the local authority or steps taken to eliminate or reduce the consequences of the deposit without serving a notice

(1) in order to prevent pollution of land, water or air or harm to human health it is necessary that the waste be removed/steps taken to eliminate or reduce the consequences of the deposit; or
(2) if there is no occupier of the land; or

(3)   if the occupier neither made nor knowingly permitted the deposit of the waste.

The necessary costs can be recovered from the occupier, unless the occupier neither made nor knowingly caused/permitted the deposit.

Under section 6 of the Refuse Disposal (Amenity) Act 1978, local authorities have a *power* (and not therefore a duty) to remove articles other than vehicles abandoned without lawful authority on any land in the open air. However, if the land is occupied, the authority is required to serve notice of its intention on the occupier. There is an appeals procedure, and the costs are recoverable from the person who deposited the articles.

Under section 34 of the Public Health Act 1961 the local authority has power to deal with any rubbish which is in the open air and which is seriously detrimental to the amenities of the neighbourhood. The power is effective 28 days after serving a notice on the owner. There is no power for the authority to recover their costs from the owner. For these purposes rubbish means rubble, waste paper, crockery, metal and other refuse (including organic matter) but not material accumulated in relation to business.

Under sections 215–9 of the Town and Country Planning Act 1990, if it appears to a local planning authority that the amenity of part of its area or an adjoining one is adversely affected by the condition of land in the area it may serve a notice on the owner and occupier requiring them to remedy its condition within a certain time. Again there is an appeals procedure. The owner or occupier may recover any expenses incurred in complying with the notice from anyone who caused or permitted the land to be in the condition that caused service of the notice.

Therefore, as will be seen from the above, there are various powers in relation to fly tipping, but only one duty, namely in respect of abandoned cars. The result of this is often stalemate – the public authority serves a notice on the occupier, but in the vast majority of cases where the fly tipping has been done by an unknown third party this is useless, as the occupier has a defence. Nor is there any joy for the occupier in this situation.

13.7   Waste disposal

13.7.1   Definitions

Part II of the 1990 Act deals *inter alia* with the disposal of waste on land. At present agricultural waste is excluded from the definition of waste (section 75(7)(c)). However, other regulations and statutory codes may apply to agricultural waste.

Under section 75(2) of the 1990 Act waste is defined as including

❐   Any substance which constitutes a scrap metal or an effluent or other unwanted surplus substance arising from the application of any process

❐ Any substance or article which requires to be disposed of as being broken, worn out, contaminated or otherwise spoilt

There is also a presumption that anything which is discarded or otherwise dealt with as if it were waste (including directive waste) is waste, unless the contrary is proved (section 75(3)).

This definition was extended by the Waste Management Licensing Regulations 1994 (WMLR) (SI 1994/1056) to correspond with the EC framework directive on waste, so that any reference to 'waste' in Part I of the 1974 Control of Pollution Act or Part II of the 1990 Act extends to 'directive waste'. Essentially, the purpose of the directive is to treat as waste, and accordingly to supervise the collection, transport, storage and recovery of objects which fall out of the commercial cycle or the chain of utility.

Waste is construed from the point of view of the discarder, whether or not it may be of value to another person. DoE Circular 11/94 and Welsh Office 26/94 on the Waste Management Regulations give practical advice on how to decide in any case if particular material is or is not waste. This identifies the following four broad categories of potential waste:

(1) Worn but functioning substances or options which are still usable (albeit after repair) for the purpose for which they were made – this is generally *not* waste.
(2) Substances or objects which can be put to immediate use otherwise than by a specialised waste recovery establishment undertaking – again these will *generally* not be waste. Examples of this might include by-products from food and drinks processing, which can be transferred into other food or drink products, or used as animal feed.
(3) Degenerating substances or objects which can be put to use only by establishments or undertakings specialising in waste recovery. These are generally considered as waste, even when they are transferred for value.
(4) Substances or objects which the holder does not want and which he has to pay to have taken away. If the substance or object has been consigned to the process of waste collection then it will be waste. If, however, it is fit for use in its present form (albeit after repair) and the holder 'intends his possession of it to determine its continued use by another identified person' the payment does not necessarily mean that it is waste. If the substance can be put to immediate use otherwise than by a specialised recovery operation, it should not normally be registered as being discarded and therefore not waste.

The 1990 Act provides that household, industrial and commercial waste is controlled waste for the purposes of the Act and includes the following (section 75):

(1) *Household waste* means waste from domestic property, a caravan, a residential home, premises forming part of a university or a school or

other educational establishment, premises forming part of a hospital or nursing home.
(2) *Industrial waste* means any waste from any factory, any premises used for the purposes of or in connection with the provision to the public of transport services by land water or air, premises used for the purposes of or in connection with public utilities, or any premises used in connection with postal or telecommunication services.
(3) *Commercial waste* means waste from premises used wholly or mainly for the purposes of a trade or business or for the purposes of sport recreation or entertainment, excluding household waste industrial waste and waste from any mine or quarry or waste from premises used for agriculture (section 75(7)).

The Secretary of State may make regulations to extend and provide for other forms of waste to be treated as being or not being household waste, industrial waste or commercial waste (as was used to extend the definition of waste so as to bring it in line with 'directive waste').

Toxic and dangerous waste is covered by the special waste provisions contained under section 62.

13.7.2 Exemptions from WMLR

There are 43 processes exempted from the WMLR. Nevertheless, in respect of most of the exempt activities, it is still necessary to register the exempt activity with the appropriate authority setting out the details as to when and where and in what quantity material is to be disposed of.

The spreading of any of the following wastes on land which is used for agriculture is exempt:

- Waste soil or compost
- Waste wood, bark or other plant matter
- Waste food, drink or materials used in or resulting from the preparation of food and drink
- Blood and gut contents from abattoirs
- Waste lime
- Lime sludge from cement manufacture or gas processing
- Waste gypsum
- Paper waste sludge, waste paper and de-inked paper pulp
- Dredgings from any inland waters
- Textile waste
- Septic tank sludge
- Sludge from biological treatment plants
- Waste hair and effluent treatment sludge from a tannery

However, the exemptions only apply if

(1) no more than 250 tonnes or in the case of inland water dredgings, 5000 tonnes, of waste per hectare are spread on land in any period of 12 months;

(2) the activity in question results in benefit to agriculture or ecological improvement; and
(3) where the waste is to be spread on land used for an agricultural undertaking or establishment it is a requirement that particulars be furnished to the waste regulation authority in whose area the spreading is to take place.

Exemption 8 exempts the storage in a secure container or a lagoon (or if unwatered in a secured place) of sludge to be used in accordance with the Sludge (Used in Agriculture) Regulations 1989, and the spreading of sludge on non-agricultural land if it results in ecological improvement and does not cause excess concentration in the soil specified elements. Both of these require registration.

Other exemptions to the WMLR relate to the recovery of wood sawdust and the keeping or treatment of animal by-products under the Animal By-Products Order 1999 (SI 1999/646).

Notwithstanding the flexible definition of waste under the guidance there are a number of notable examples where waste management licences would be required, e.g. the use of hardcore supplied by a road building company to a farm to fill in potholes or a gateway would require a waste management licence. If, however, hardcore had been produced elsewhere on the farm and used on it no licence is required. Similarly waste used for ecological improvement or reclamation of land is only exempt with the requisite planning permission.

13.7.3 *The waste management licence*

A *waste management licence* authorises the treatment, keeping or disposal of any specified description of controlled waste in or on specified land or the treatment or disposal of any description of controlled waste by means of mobile plant (section 35 1990 Act). The licence is granted to the occupier, where the licence is for the treatment, keeping or disposal of waste in or on land or the operator where the licence is for the use of mobile plant to treat or dispose of waste. For a licence to be granted, three requirements must be met

(1) Planning permission is in force or an establishment use certificate is in force
(2) The applicant is a fit and proper person (see section 74 for detail on this)
(3) The refusal is necessary for the purposes of preventing pollution of the environment, harm to human health or serious detriment to the amenities of the locality where there is no planning permission in force

Where the authorities are minded to grant a licence, the proposal should be referred to the Environment Agency and the Health and Safety Executive and representations from both bodies should be considered. If the licence is not forthcoming within four months from the date of application, then the application will be deemed to have been rejected (section 36 1990 Act).

Once granted, the conditions of the licence may be modified whilst it is still in force, or even revoked in certain circumstances (see sections 37–39).

There is a right of appeal to the Secretary of State against the decision of a waste regulation authority on applications, conditions, variations, suspensions and revocations, etc. The Secretary of State may appoint someone to hear the appeal and it may be determined in public or private (section 43).

13.7.4 Offences

It is an offence to deposit waste except in accordance with a waste management licence, or otherwise to breach the WMLR. Conviction can lead to a fine of up to £20 000 on summary conviction or on indictment an unlimited fine, and five years imprisonment in relation to special waste, two years in relation to controlled waste. Notwithstanding the regime, regard should also be had to the potential hazards of using land for the disposal of waste in terms of the contaminated land regime and potential liability that may arise from European legislation such as the environmental liability directive and the product liability directive.

A person charged with an offence under the 1990 Act has three defences:

(1) he took all reasonable precautions and exercised all due diligence to avoid the commission of the offence; or
(2) he acted under instructions from his employer and neither knew nor had reason to suppose that the acts done by him constituted an offence; or
(3) that the acts alleged to constitute contravention were done in an emergency in order to avoid danger to the public and that, as soon as was reasonably practicable after the acts were committed, particulars of them were furnished to the waste regulation authority in whose area the treatment to dispose of the waste took place.

13.7.5 The duty of care as respects waste

Under section 34 a duty is imposed on any person who imports, produces, carries, keeps, treats or disposes of controlled waste or as a broker has control of such waste to take all such measures applicable to him in that capacity as are reasonable in the circumstances.

This duty does not apply to an occupier of domestic property when he is dealing with household waste produced on his property.

Effectively, therefore, the duty falls on any person concerned with controlled waste, and the maximum penalty for breach of the duty is an unlimited fine.

It is an offence to transport controlled waste in the course of a business or for profit without being registered. There are various defences to this, such as emergency, or acting on the instruction of one's employer.

13.9 The Hedgerow Regulations 1997

The Hedgerow Regulations 1997 (SI 1997/1160) introduce restrictions on the removal of hedgerows. The basic scheme is as follows. An owner wishing to remove a hedgerow must notify the local planning authority in advance. The authority will assess the importance of the hedgerow against certain criteria and decide within six weeks whether to oppose the removal of the hedgerow. If the authority finds that the hedgerow is not important, the owner will be able to remove it. If the hedgerow satisfies one or more of the criteria, the authority will then decide whether or not to issue a hedgerow retention notice, having regard to the reasons put forward by the owner for wanting to remove the hedgerow. The authority may decide that the reasons for the removal override the importance of the hedgerow: it can then be removed. If the authority issues a retention notice, the owner can appeal.

This protection only applies to hedgerows which are at least 20 m long, or if shorter, adjoined to other hedgerows at both ends. A hedgerow need not be solid throughout its length: gaps of up to 20 m will be counted as part of the length of a hedgerow. Hedgerows less than 30 years old will be exempted from notification. Notification will not be required where hedgerow removal has been sanctioned as part of a grant of planning permission. Also, management work (trimming, coppicing and laying) is excluded from any requirement for notification. The eight criteria which local authorities must apply to determine whether a hedgerow is 'important' are if the hedgerow in question

- Marks a pre-1850 parish or township boundary
- Incorporates an archaeological feature
- Is part of, or associated with, an archaeological site
- Marks a boundary of or is associated with a pre-1600 estate or manor
- Forms an integral part of a pre-parliamentary enclosure field system
- Contains certain categories of species of birds animals or plants (this is contained in the Wildlife and Countryside Act)
- Includes a certain number of woody species
- Runs alongside a public bridleway, footpath, road used as a public path or a byway open to all traffic, and includes at least four woody species

Breach of the Hedgerow Regulations is a criminal offence, and can lead to an unlimited fine. Further in the event of the removal of a hedgerow, or part thereof, in breach of the Regulations, the authority can require the owner to reinstate the hedgerow.

Chapter 14
Milk Quotas

14.1 Introduction

Milk quotas were introduced by Council Regulation (EEC) No 857/84 on 2 April 1984. Each member state was allocated a guaranteed total quantity of quota. Having received its national quantity, it was then up to member states to allocate quota on an individual basis. The UK chose to base the allocation on the 1983 level of production, but allowed producers to opt for an earlier year in specified circumstances.

Milk quotas were originally introduced for a five-year period but, following repeated extensions, they are currently (2002) set to continue until 2008. Since their creation various amendments have been made to the European and UK legislation to address particular problems and to provide greater flexibility. The present law is set out in Council Regulation (EEC) No 3590/92 as amended, Commission Regulation (EEC) No 536/93 as amended and the Dairy Produce Quotas Regulations 1997 (SI 1997 No 733) as amended. The quota system is now administered by the Rural Payments Agency (RPA), which is responsible for the maintenance of producer registers.

14.2 The milk quota system

Quota is most easily defined as the quantity of dairy produce which may be sold from a holding without the liability to pay a levy. England and Wales opted at the outset for payment of levy under what was then known as Formula B. This meant that the levy had to be paid by the purchaser of milk on amounts delivered to the purchaser in excess of quota. The Council Regulation changed this structure but member states still have the option to elect for payment of levy at purchaser or at national level. The UK has chosen the purchaser level in line with its original choice of Formula B. Following the revocation of the milk marketing schemes in 1994, the five milk marketing boards, which until then fulfilled the role of purchaser, were replaced by a variety of purchasers, with whom producers can negotiate individual contracts to supply milk and milk products.

There are two different types of milk quota: *direct sales quota* is quota allocated to a producer who sells milk direct to his customers; and *wholesale milk quota* is quota allocated to a producer who sells milk to a purchaser, who then processes and distributes the milk and milk products. In the case of wholesale milk supplies, the producer receives payment for his milk under the terms of his contract with his chosen milk purchaser and the producer's quota effectively forms part of that purchaser's quota for the purposes of calculation of the levy. The regulations provide opportunities for a producer to convert his quota either from wholesale to direct sales quota or vice versa on a temporary or on a permanent basis.

Following the introduction of quotas the butterfat content of milk continued to rise, creating an ever greater dairy surplus. In response to this, producers and purchasers were given a fixed butterfat base. This means that for the purposes of calculating the levy each year, deliveries of milk with an average fat content in excess of or below the butterfat base will result in an adjustment of the amount of milk delivered.

The milk quota year runs from 1 April in one year to 31 March in the next year.

In 1994, a National Reserve of milk quota was established by Council Regulation (EEC) 3950/94 for the holding of quota, which was not allocated to any producer. This included quota confiscated under that regulation due to non-use or due to a breach of restrictions imposed in relation to special quota (see section 14.3 below). The regulation provides a framework for the restoration of confiscated quota in certain circumstances, although this is at the discretion of the RPA, rather than being an absolute right of the producer. To avoid confiscation producers should therefore take steps each milk year either to produce against their quota or to lease at least a part of it to another producer.

14.3 Special quota

Allocations of special quota were made after 1984 to former milk producers who had either adopted milk production development plans, or who had gone out of milk production under one of the outgoing milk schemes before the introduction of milk quotas. Quota allocated in those cases is known as *SLOM quota*. Various restrictions were imposed on the allocation of special quota, some of which were challenged successfully in the European Court and which resulted in further allocations of special quota. Of particular note were the restrictions imposed on the transfer of special quota, although the time limits imposed have now all expired.

14.4 Transfer of quota

14.4.1 With land
Transfer of milk quota between producers has over the years become one of the most important issues arising out of the milk quota system. From the outset, it was intended that quota would be linked to the land to which it was allocated and that it would be transferred with that land. The regulations have always provided for the transfer of quota when any holding is sold, leased or transferred by inheritance.

A *holding* within the meaning of the regulations can be paraphrased as being all the land occupied by a registered milk producer at any one time. A registered producer's holding will often comprise both land which he owns freehold and land which he occupies as a tenant.

Where only part of the holding is transferred, an apportionment of quota has to be made between the part of the holding sold or leased out and the part retained by the transferor. The apportionment can be carried out either by agreement between the parties or by arbitration. It should take account of the areas used for milk production by the transferor and where an arbitrator makes an award he will look at the areas so used for the five years preceding the change of occupation. If the apportionment is made by agreement, the parties decide how many litres of quota should go with the land sold or leased. If the RPA has reasonable grounds for believing that the apportionment does not take account of areas used for milk production, then the board may serve notice to this effect, whereupon an arbitrator can be appointed under the DPQR 1997 to settle the apportionment.

There is no definition of the phrase 'areas used for milk production' but it was given a wide interpretation in *Punchknowle Farms Ltd* v. *Kane* [1985] 3 All ER 790. That case decided that areas used for milk production included

> 'The area of a set of buildings and yards belonging thereto used for the production of milk and the forage areas used by the dairy herd and to support the dairy herd by the growing of grass and any fodder crop for the milking dairy herd, dry cows and all dairy following female youngstock (and home bred dairy or dual purpose bulls for use on the premises, if applicable) if bred to enter the production herd and not for sale. In this case, maize, silage, hay and grass were the fodder crops, but consideration would have been given to corn crops, or part of corn crops, growing for consumption by the dairy herd or youngstock, including the use of straw, had agreed evidence been produced.'

Where the parties themselves are unable to agree the areas used for milk production and the matter goes to arbitration, the arbitrator has to consider the historical evidence for the five years immediately preceding the change of occupation. It is generally considered as a result of this case that quota must be apportioned on an acre for acre basis (the mathematical method) rather than on the basis of the productivity of the land and fixed equipment. The ministerial guideline is that no more than 20 000 l per hectare should be transferred.

In case the amount of quota is subsequently queried by the RPA, parties often negotiate the inclusion within the contract or lease of a provision for compensation at the market value to be paid to the party suffering loss, should the agreed apportionment be challenged at a later stage. It is possible for a producer to obtain a prospective apportionment of quota which will be binding if a transfer is made within six months of the apportionment.

The transfer of quota has to be registered with the RPA using prescribed forms which must be submitted by the earlier of a date 28 days from the change of occupation of the holding or by 31 March (1 March for transfers of quota with leases of land). The form must be signed by both transferor and transferee and must include a 'consent or sole interest notice' in respect of the holding. This notice either confirms that there is no other party apart from the transferee who has an interest in the holding or that all persons having an interest in the holding, the value of which interest might be reduced by the apportionment of the quota, agree to that apportionment. In practice all parties with an interest in the holding must sign Part B of the Form to confirm their consent. There is no definition of interested parties but they will include a landlord, mortgagee, trustee, beneficiary, a party with an option over the land and a party which has contracted to purchase any part of the holding. On a transfer of quota both the transferor and transferee must notify their respective milk purchasers within seven days of the transfer.

There was initially considerable uncertainty concerning which transactions, other than a disposal of the freehold or part of the freehold, would enable milk quota to be transferred. The position was clarified in 1988 and the current provisions are set out at Article 7 of DPQR 1997. They provide that in England and Wales, no person shall transfer quota on a transfer of the holding or part of a holding in the following cases:

(1) A licence granted to occupy land
(2) A tenancy granted for less than ten months (in England and Wales)
(3) The termination of a tenancy granted for less than 10 months
(4) The termination of a licence

Under both the European and UK legislation quota was initially only intended to be transferred together with the land to which it was allocated. In the UK, however, a transfer system developed which was based on the granting of short-term tenancies. If a producer wants to sell part of his quota without selling his land he can enter into a quota sale agreement with another producer, under which the buyer takes a short-term tenancy of a part of the seller's holding and occupies that land for non-milk producing purposes. In England and Wales the tenancy must be for a period exceeding 10 months. The quota is transferred at the start of the tenancy. At the end of the term the quota (then registered with the buyer) attaches to the areas, which the buyer has used for milk production, which does not include the tenanted land. In this way the quota is transferred to the buyer.

Where the buyer lives in another part of the country it may be necessary to enter into a further agreement for the property to be farmed by an independent contractor. The independent contractor will often be the seller. However, for there to be the necessary change of occupation to enable the milk quota to pass to the buyer, it is essential that the seller farms as agent for the buyer and not as owner. Moreover, during the term of the tenancy the land which is the subject of the tenancy must not be used for milk production.

With the introduction of the Integrated Administration and Control System (IACS) in 1993 together with the Arable Area Payment Scheme and the various other livestock subsidy schemes governed by IACS, particular issues can arise on the transfer of milk quota in this way. This is especially if the parties wish to claim arable area payments for, or to register as forage area, the land the subject of the short-term tenancy. For the purposes of the milk quota transfer it is clearly the buyer of the quota who must be in occupation of the land and it is therefore he, rather than any contractor, who is entitled to include this land on his IACS form.

14.4.2 Without-land transfers

In 1992 the Council Regulation introduced the opportunity for permanent transfers of quota without land. The UK took up this opportunity and the details of this method of transfer are set out in the DPQR 1997. Essentially, without-land transfers are permitted where there is an aim of improving the structure of milk production at the level of the holding. An application must be made by the transferee to the RPA no later than 10 days before the intended date of transfer. The application must contain an explanation by both transferee and transferor as to how the transfer of the quota will improve the structure of both businesses.

A consent or sole interest notice must also be signed by the transferor (as explained above) and both parties must give certain undertakings regarding the transfer of quota to and from their respective holdings, both before and after the transfer. In particular, for the remainder of the year of the transfer and the following quota year, the transferor is restricted from buying or leasing in any quota and the transferee is restricted from selling or leasing out any quota. The RPA has discretion to waive these restrictions in certain exceptional circumstances. If the RPA accepts the proposal it will authorise the transfer of quota and will set the actual date of transfer. The RPA will not generally give any assurance in advance that a particular transfer will be accepted. This has made the use of without-land transfers difficult where the seller is giving up occupation of his land at the same time as selling his quota.

14.4.3 Particular questions

Several questions are frequently asked concerning the transfer of milk quota. The first one is whether the grant of a grazing agreement will trigger a transfer of quota. The answer is that quota will only be transferred under the regulations if the agreement amounts to a tenancy, i.e. it gives the grazier exclusive possession and is for a term of over ten months. Since 1 September 1995 such grazing tenancies are likely to be farm business tenancies if the tenant uses the land for the purposes of a trade or business. If the parties do not intend the quota to transfer to the tenant then, as this does not accord with the regulations, specific provisions should be included in the grazing tenancy agreement to minimise the risks of a strict enforcement of the regulations.

The second question is whether a landlord can prevent a tenant from transferring quota by means of a subtenancy. The answer is that this depends on the terms of the head-tenancy agreement. Every well drawn tenancy agreement should have a covenant against subletting or assignment. Where there is no such clause and the tenancy is governed by the Agricultural Holdings Act 1986 (AHA), landlords can have a clause written in as provided by section 6 of the AHA. In any event, a landlord has an interest in the holding of the tenant and for the transfer to be valid the tenant needs the landlord to sign the consent part of the transfer form.

The third question concerns 'massaging' by a tenant. A tenant may be a registered milk producer in respect of tenanted land and also other land which he owns freehold (or rents from another landlord). He could, therefore, remove his dairying activities from his let land to his other land so that after a period of time on the termination of his tenancy the let land is not used for milk production. No milk quota would then revert to the landlord. Can the landlord prevent this? The answer depends on the terms of the tenancy agreement. Where agreements have been entered into since the introduction of quotas there should be provisions in the agreement to ensure that milk quota is preserved for the benefit of the let land. The position is more complex where there are pre-1984 agreements. A landlord might be protected if there are conditions in the agreement about maintaining quotas generally or if there is a provision that the farm should be preserved as a dairy farm – but not otherwise.

The fourth question concerns share farming agreements. Does a farmer with milk quota, who enters into a share farming agreement with a landowner, whereby they share an interest in cattle or crops, endanger his milk quota because the landowner may claim some of it? The answer to this question is no. In a properly drawn share-farming agreement there is no tenancy and therefore no change of occupation. The landowner retains occupation of his own land; the farmer enters the other's land for strictly limited purposes. There is no change of occupation sufficient to activate a transfer of quota.

14.4.4 The Nature of milk quota

As mentioned above, it was intended from the outset that milk quota would be linked to the land to which it was allocated. The exact nature of that link and the legal status of quota, however, still remains unclear, particularly in light of the fact that quota can sometimes be transferred under the regulations without the granting of any interest in land.

The nature of the link with land was considered by the High Court in the case of *Faulks* v. *Faulks* [1992] 15 EG 82. Chadwick J accepted, that, although milk quota had an intrinsic value, it was still linked to the holding and was not an asset which was separate and distinct from the holding.

This case can be contrasted with the later tax case of *Cottle* v. *Coldicott* [1995] SpC 40 in which the Special Commissioners decided that for the purposes of the capital gains tax legislation milk quota constituted a separate asset when it was transferred by a short-term tenancy.

This issue was further addressed in the context of the interest of a mortgagee in the case of *Harries* v. *Barclays Bank plc* [1997] 2 All ER 15. The judge agreed with Chadwick J that milk quota is not a right or interest in property which is capable of being separated from land.

In light of the continuing uncertainty regarding the nature of milk quota, it is vital that all parties with an interest in the holding to which the quota is allocated, such as a mortgagee or a partner in a partnership, should consider their position carefully particularly if any transaction relating to the land or any change in the occupation of the land is proposed.

14.5 Leasing of milk quota

The DPQR 1997 permit a producer, including a tenant, to effect a temporary transfer of the whole or any part of his quota, which he does not need to use during a quota year to another producer who requires additional quota. These transfers are commonly known as quota leases. They are effected by the completion by both parties of the prescribed RPA form, which must be submitted to the Board by 31 March. Once the lease is accepted by the RPA, the leased quota is available to the lessee for the quota year of the lease and the quota then reverts back to the lessor at the start of the following quota year.

Although the leasing forms indicate that landlords should be informed of the completion of any lease, there is no statutory requirement for any landlord to consent to the lease. Tenants might, however, have to obtain consent if a covenant to that effect is included in the tenancy agreement.

14.6 Milk quota and tenancies

14.6.1 *Agricultural Holdings Act 1986 or Agricultural Tenancies Act 1995*

Various milk quota issues arise where land is let out for the purposes of milk production. These issues differ depending on whether the letting is governed by the Agricultural Holdings Act 1986 (AHA) (essentially tenancies granted before 1 September 1995) or whether the letting is a farm business tenancy under the Agricultural Tenancies Act 1995 (ATA).

14.6.2 Rent

Tenancies governed by the AHA are subject to periodic rent reviews. In arriving at a figure for rent various factors have to be taken into account including 'the productive capacity of the holding and its related earning capacity'. If an occupier of an agricultural holding has quota registered in his name it clearly affects the agricultural holding's earning capacity.

The AHA provides that in fixing the rent an arbitrator shall disregard any increase in the rent value due to tenant's improvements. The definition of tenant's improvements in the AHA requires the improvement to have been executed on the holding wholly or partly at the expense of the tenant. Allocated milk quota does not fulfil this definition and is not therefore a tenant's improvement.

One category of milk quota, however, can be disregarded in assessing the rent. Section 15 of the Agriculture Act 1986 (AA) directs an arbitrator to disregard quota which was transferred to the tenant by virtue of a transaction, the cost of which was borne wholly or partly by him. There are several difficulties with this phrase. 'Transaction' is not defined but it is generally taken to mean at least a formal contract. Moreover, it is not clear what the position would be where the cost is paid not by the tenant but by the partnership or company in which the tenant has an interest.

There is a further problem in that any payment made by the tenant for the grant or assignment of a tenancy with the benefit of quota does not count as quota transferred at the tenant's cost. There are differing views as to whether this covers a payment by a new tenant to a landlord who requires to be compensated for the milk quota compensation he has paid to an outgoing tenant.

It is possible to contract out of section 15 AA 1986. Therefore, on the grant of a new tenancy, a landlord, who wants to maintain the rent level, should obtain the agreement of the prospective tenant, that any milk quota paid for by the tenant should not be disregarded in assessing the rent. This agreement should be recorded in the written tenancy agreement.

The position under the ATA for farm business tenancies is rather different. The parties have greater freedom to negotiate the timing and the form of rent review, which will apply, although where the arbitration procedure is chosen, the ATA prescribes those matters which are to be disregarded and those which are to be taken into account. If milk quota constitutes a 'tenant's improvement' (as defined by the ATA), it is generally disregarded, whilst all other quota will be taken into account. There are certain exceptions to this general rule. The definition of tenant's improvement includes 'any intangible advantage which – (i) is obtained for the holding by the tenant by his own effort or wholly or partly at his own expense, and (ii) becomes attached to the holding'. This definition is generally accepted to include milk quota where the tenant fulfils the conditions. If rent review by arbitration is chosen then contracting out of these provisions is prohibited by the ATA.

14.6.3 Compensation

14.6.3.1 The Agriculture Act 1986

The AA 1986 provides for certain tenants of tenancies governed by the AHA to obtain compensation for milk quota on the termination of their tenancy.

Eligibility is covered in paras 1–4 of Schedule 1 AA 1986. An eligible tenant is a person who has milk quota registered in his name in relation to land and in addition falls into one of the following categories:

(1) A tenant who had milk quota allocated to him or was in occupation on 2 April 1984 and had milk quota subsequently transferred to him wholly or partly at his cost.
(2) Statutory successors on the death or retirement of such a tenant.
(3) Assignees of such a tenant whether by deed or operation of the law.
(4) Tenants who have sublet, where the subtenancy terminated after 2 April 1984, and the subtenant is entitled to claim compensation from him.

There is some difficulty over partnerships and companies because AA 1986 only allows compensation where the milk quota is 'registered as his'. For partnerships it is generally the firm, and not the individual members of the firm, whose name appears on the RPA's register. Logically the position would seem to be as follows:

(1) If the tenant is not a member of the partnership which is registered with milk quota then he cannot claim.
(2) If he is a member of the partnership then he can claim the full amount of the compensation. He should then distribute the compensation in accordance with the terms of the partnership agreement.
(3) If the tenant is an individual but the milk quota is registered in the name of a limited company, then there is unlikely to be a claim to compensation. A company is a distinct legal entity and even the tenant's involvement as a director or shareholder of the company would not alter the position.
(4) If the tenant can show that the company held the registered milk quota on trust for him then he will be entitled to compensation.

No compensation is payable to the retiring tenant or to the deceased tenant's estate when a statutory succession takes place. Similarly, in the case of assignments and subtenancies the right to compensation passes to the new tenant. In other words, the legislation gives statutory entitlement to compensation on only one transfer of occupation.

Relevant quota is dealt with by para 1, Schedule 1 AA 1986. In cases where the land, which is being vacated on termination of the tenancy comprises all the land occupied by the tenant, then all of the quota registered in the name of the tenant will be relevant to the calculation of compensation. The volume of quota can be established from the RPA. In those cases where the tenant occupied other land then it will be necessary to apportion the allocated quota between the land being vacated on termination of the tenancy and other land. Apportionment of a quota depends on the areas used for milk production. Procedures for apportionment are set out in the DPQR 1997.

Paragraph 10 of Schedule 10 AA 1986 covers the determination of *standard quota* and *tenant's fraction* before the end of the tenancy. Liability for compensation does not arise until the termination of a tenancy. However, the landlord or tenant may at any time before the termination of the tenancy seek to determine the amount of the standard quota or the tenant's fraction. This may be done by agreement between the parties, or failing that by reference to arbitration. Arbitration has to be initiated by notice in writing by either party. The arbitrator can determine the amount of standard quota and the tenant's fraction before the termination of the tenancy. However, this determination is only valid if there are no changes in government regulations or in the amount of relevant quota. The value can only be settled after termination.

Paragraph of Schedule 1 of the AA 1986 provides for calculation of the *value* of the tenant's compensation as follows:

> '... the value of milk quota ... shall take into account such evidence as is available including evidence as to the sums being paid for interests in land –
> (a) in cases where milk quota is registered in relation to the land and
> (b) in cases where no milk quota is so registered.'

Most of the evidence available in respect of sales of quota is based on small litreage and this could justify arguments that the value of larger litreages may be for lesser sums per litre. The value of quota may vary considerably over the years due to the scarcity and profitability of milk production and the European Community rules on leasing and transfer of quota.

Settlement of a tenant's claim on termination of his tenancy is covered by paragraph 11 of Schedule AA 1986. The tenant is required to make a claim for compensation against his landlord within two months of the end of the tenancy. If the parties cannot agree, settlement will be by arbitration.

14.6.3.2 *The Agricultural Tenancies Act 1995*

The position regarding compensation for tenant's improvements on termination of a farm business tenancy is quite different from that which applies to AHA tenancies. First, the ATA specifically provides that the AA 1986 shall not apply to farm business tenancies. Instead, compensation is provided under the ATA for those items which constitute 'tenant's improvements' under section 15 of the ATA (see section 14.6.2 for definition).

It is generally accepted that the definition of improvements includes milk quota, however, for a claim for compensation to succeed the tenant must also obtain the landlord's consent in writing to the improvement. This issue is best addressed in the tenancy agreement, but if it is not and if the landlord refuses consent, then the ATA does in certain circumstances allow the tenant to refer such a refusal or the imposition by the landlord of unreasonable conditions to arbitration under the ATA.

The calculation of compensation under the ATA is also different from the AA 1986. Under the ATA the amount due is a sum 'equal to the increase attributable to the improvement in the value of the holding at the termination of the tenancy as land comprised in a tenancy'. How the level of this compensation compares with that available under the AA 1986 will very much depend on the value of quota at the time of the termination of the tenancy.

Although the ATA provides for compensation to be rolled over between successive farm business tenancies, particular care should be taken where a tenant moves, by agreement with a landlord, from a tenancy protected under the AHA to a farm business tenancy. In those circumstances, compensation for the quota might well be available under the AA 1986 on termination of the first tenancy. Problems will arise, however, if the parties wish to roll over the compensation into the farm business tenancy. First, the quota might not fulfil the definition of tenant's improvement and, second, the value of the compensation might be less under the ATA.

In view of these issues it is even more important under the farm business tenancy regime that the parties address and provide in the tenancy agreement for compensation for milk quota where this is appropriate.

Chapter 15
Employment

15.1 Introduction

When an employer employs a worker he enters into a contract. This contract will specify the job which the worker has agreed to do and the wages he or she will receive for the work. It will also cover any other agreed terms. These are likely to include provisions for holidays, sick pay and notice for termination of the employment.

If there is a breach of these contractual terms which arise on or are outstanding on termination of employment the *Employment Tribunal* has jurisdiction to hear such an action (up to a maximum of £25 000) since 1994. Redress for civil wrongs not involving contract, for example claims for personal injury, can be brought in the civil courts in the normal way. In addition to these common law rights and remedies workers are given rights under employment legislation. These rights cannot be removed or reduced by any contractual provisions. On the other hand, employees can be given additional rights by their contract.

15.2 Employment Tribunals

Claims to enforce the statutory employment rights are dealt with by Employment Tribunals. An Employment Tribunal consists of a legally qualified chairperson and two lay members with industrial experience. One lay member will be selected from a panel representing employers and the other from a panel representing employees. Tribunals sit in different parts of the country.

The intention behind tribunals is that they should provide a cheap, quick and informal method of settling employment disputes. An individual worker can begin tribunal proceedings by filling in a simple standard form available from job centres and unemployment benefit offices.

An employer, or employee, may represent himself before the tribunal, or he may be legally represented, or he may be represented by some other person. An employer might be represented by his farm manager or agent, an employee by a trade union official.

Rather than spend time on attending a hearing an employer faced with a complaint may try to settle the case. Any agreement reached will be binding on a worker only if reached with the assistance of an ACAS conciliation officer. ACAS (the Advisory, Conciliation and Arbitration Service) is an independent body which was set up to improve industrial relations. ACAS conciliation officers are sent copies of complaints presented to Employment Tribunals. They are then required to try to promote a settlement at the request of both parties or on their own initiative if they think a settlement is possible. As mentioned, if the parties reach a settlement themselves, they must call in a conciliation officer to make it valid. If the case later goes to a tribunal, nothing said to the conciliation officer is admissible in evidence without the consent of the person who said it.

Most cases are heard by tribunals within 16 weeks of an application being made. There are statutory time limits, depending on the particular cause of complaint, for making an application to a tribunal. In some circumstances the tribunal will extend the period where it was not reasonably practicable for the complainant to comply with the time limit.

It is possible to make an appeal on a point of law to the Employment Appeal Tribunal (EAT). Generally, there can be no appeal against the tribunal's finding of fact. The EAT consists of a judge and two to four lay members selected for their experience in industrial relations. Further appeals can be made, with leave of a court, to the Court of Appeal and the House of Lords. Most appeals do not go beyond the EAT.

A new scheme was introduced recently by ACAS in May 2001 to provide an alternative to Employment Tribunals in the case of unfair dismissal. There will be a voluntary confidential arbitration as a means of resolving unfair dismissal claims. There will not be an appeals process.

15.3 Employees or self-employed workers

15.3.1 *The consequences*

It is important to decide whether a worker is an employee or a self-employed worker. Certain consequences follow from being an employee which do not arise where the worker is an independent contractor. These are

(1) Many employment rights apply only to employees. These include the right to complain of unfair dismissal, to claim statutory redundancy payments, to have guaranteed pay and specified time off, and to have a written statement of employment terms. Increasingly new rights are being introduced which apply more generally to 'workers'.

(2) Income tax payable by employees is deducted at source by employers on the Pay As You Earn basis. The self-employed are paid for their services gross and are liable to pay their own tax out of the money they receive.

(3) Employees are liable for Class 1 social security contributions, part of which is paid by their employer. The self-employed pay their own contributions, at a lower rate. Only employees can claim unemployment benefit.

(4) Where an employee commits a civil wrong in the course of his

employment, his employer will be vicariously liable to the injured third party. The injured party can sue the employer for damages as well as, or instead of, the employee. Employers should, of course, insure themselves against such risks. In some cases, such as injuries caused to fellow employees and road traffic accidents, insurance is compulsory. Employers are not, in general, responsible for the wrongful acts of the self-employed.

(5) Some industrial safety legislation applies only to employees and they are owed higher duties than the self-employed under the Health and Safety at Work Act 1974. The standard of care owed at common law to employees is also higher.

(6) If an employer goes bankrupt, employees are preferred creditors in respect of any wages owed to them. They can also apply for payments to the Redundancy Payments Office of the Department of Trade and Industry. No preference is given to the self-employed.

15.3.2 Deciding on status

It is not an easy matter to decide whether a worker is an employee or an independent contractor. It depends on all the circumstances. Because of the matters listed above an employer may decide it would be more advantageous to call the employee self-employed. However, the Inland Revenue, the Benefits Agency, the Employment Tribunal, and the Courts will disregard any label if it does not accord with the true facts.

Set out below are some questions which will be considered when deciding on the status of the worker.

(1) Does the worker work a regular 35-hour week for one employer at a fixed wage? If so, he will be an employee. On the other hand a hedge-cutter who works for a number of people and only comes a few times a year is likely to be treated as self-employed.

(2) How much control does the employer have over the worker? If the employer can tell the farm worker not only what work to do but also how to do it the worker will probably be treated as an employee. The converse, however, is not necessarily true. An employee with specialised skills would expect to use his own judgement and not be dictated to. Yet he could still be an employee. Another factor to be taken into account is whether the worker's activities are an integral part of the employer's business or only accessory to it. In these days of diversification a worker may in fact be an entrepreneur running his own business alongside that of the farmer. In this situation he would be treated as self-employed. This will be especially so if he provides his own equipment.

(3) Does the worker work for only one employer? If he does this may indicate he is an employee. He may also be an employee if he works

part-time on a regular basis for one employer even where he also works for others. On the other hand, employee status does require both sides to be subject to some degree of obligation. This would exclude entirely 'casual' work.

15.3.3 Directors

The question is sometimes asked as to whether a director is an employee of a company. This again depends on the circumstances. All directors have duties and responsibilities towards the company but not all directors are employees.

A director may have a formal contract of employment with a company which constitutes him an employee. Abuses used to be possible because company directors obtained long-service contracts with their companies. If they were then dismissed they could claim substantial compensation for breach of contract. Section 319 of the Companies Act 1985 now provides that a company cannot enter into a service contract for more than five years with a director, unless the company has approved the contract in a general meeting. Contracts for more than five years which have not been so approved are deemed determinable by the company on giving reasonable notice, at any time, to the director. The director will not be entitled to compensation even if his contract is for a term in excess of five years.

Even without an express contract it is possible to imply one especially where a director 'devotes his whole time to the affairs of the company; does all in his power to develop and extend the company and does not engage in any other business'. Matters to be taken into consideration when determining the status of a director are

(1) Whether he is paid and taxed as an employee or is merely paid directors' fees. This information will be found in the company's accounts
(2) Whether his National Insurance contributions are deducted by the company or whether he pays them as a self-employed person
(3) Whether the company keeps written evidence of a service contract
(4) Whether he works consistently and regularly for the company

15.4 Engaging a worker

15.4.1 Discrimination

A landowner or farmer who engages a worker, either as an employee or as an independent contractor, is subject to the provisions of the Sex Discrimination Acts 1975 and 1986, the Race Relations Act 1976 and the Disability Discrimination Act 1995. These Acts make it unlawful for employers to discriminate against job applicants and workers because of their sex, race, marital status or disability. Discrimination can arise in advertisements for jobs, in the selection procedure, and in the way promotions are made. It is unlawful to deny workers training or other benefits or to choose workers for dismissal because of their sex, race or disability.

Workers must not be paid less because of their race nor must they be given less favourable terms than other groups. The Equal Pay Act 1970 requires equal pay to be given to men and women. All part-time employees now qualify for statutory employment rights on the same basis as full time employees. From 1 July the Part-Time Workers (Prevention of Less Favourable Treatment) Regulations 2000 came into force and cover not just employees but workers. These Regulations are designed to ensure part-time workers are treated on a par with full-time colleagues.

A worker who considers he is the victim of discrimination can refer the matter to an Employment Tribunal. There is no upper limit to the award a tribunal can order an employer to pay the worker. It can also recommend that the employer takes action to remove or reduce the adverse effect of the discrimination.

15.4.2 Persons with a criminal record
A prospective employee is not bound to disclose to an employer convictions which have become 'spent' under The Rehabilitation of Offenders Act 1974. The Act does not allow sentences of imprisonment exceeding 30 months to be 'spent'. Other convictions are 'spent' only after the periods laid down in the Act. A non-custodial sentence becomes 'spent' after five years; a 30-day prison sentence after seven years. It is unlawful to exclude a rehabilitated offender from any occupation or employment because he has a spent conviction or has failed to disclose one. Nor is it lawful to dismiss a person for that reason.

15.4.3 Contract of employment
There is no legal requirement that a contract of employment should be in writing. However, employers must give their employees a written statement setting out their basic employment terms. This must be done within 13 weeks of the employment starting. Where an employee has been given a written contract his contract may contain all the statutory terms which have to be included in a written statement. If not, he will be entitled to a written statement to supplement his contract.

15.4.4 Contents of the written statement
These may seem formidable, but actually they cover basic matters which any employee would expect to be told about. They are as follows:

(1) The parties.
(2) The date employment began and the date when the employee's period of continuous employment began taking into account any employment with a previous employer which may count.
(3) Salary; wages; rates of pay, including overtime and piece work rates.

(4) Intervals when wages or salary are paid, that is whether they are paid weekly, monthly or quarterly. Every employee has a right to be given an itemised pay statement in writing, setting out the gross amount of wages or salary, the amount and purpose of deductions and the net amount of wages and salary (section 8 Employment Rights Act 1996).

(5) Terms and conditions relating to hours or work, including normal working hours. Where hours of work are flexible or there are no set hours this should be stated.

(6) The title of the worker's job should be stated. Some farm workers are employed for special work, for example as a dairyman, pigkeeper, shepherd, tractor driver, but are expected to do other farm work occasionally. If that is the case, it should be made clear in the statement of terms.

(7) Place of work.

(8) Terms and conditions relating to holidays and holiday pay. The current Agricultural Wages Order regulates holidays and holiday pay for farm workers. In the case of other workers, the employer should make clear what holidays the worker is entitled to. (The Working Time Regulations entitle workers to a minimum of four weeks paid leave per year which may include public and bank holidays.)

In addition the following clauses should be included:

❐ Terms and conditions relating to incapacity for work due to sickness or injury. Whether provision will be made for pay during sickness, and if so, for how long. The Statutory Sick Pay Scheme makes provision for employers to pay sick pay for a maximum of 28 weeks in the year, and to deduct the cost from National Insurance contributions. These payments apply to agricultural workers as well as other workers, and notes are issued by DEFRA to explain how sickness payments under the Agricultural Wages Order relate to payments under the Statutory Sick Pay Scheme.

❐ Any terms and conditions relating to pensions and pension schemes. From 8 October 2001 most employers who employ five or more people have a legal obligation to offer a stakeholder pension plan with a payroll deduction facility to its employees. Employers will not have to make contributions on behalf of the employees but many will need to ensure that an appropriate scheme is set up.

❐ Length of notice for terminating employment. This must be *at least* the minimum laid down in section 86 Employment Rights Act 1996 (see section 15.6 below).

❐ Information about grievance procedure, telling the employee to whom he should apply if he has a grievance about his employment, and how the matter will be dealt with. The procedure can be elaborate in a large establishment, but where there are only a few employees, notice of grievances in writing given to the agent, foreman or employer should suffice. Information should also be given about disciplinary rules (other than those relating to health and safety at work). The Employment Rights Act 1996 compels employers of 20 or more employees to include a note of disciplinary procedures in the written statement. Those who employ fewer are required to identify the person to whom grievances should be addressed and how they should be made.

It is not necessary to set out all this information in the statement of terms itself, provided a document or notice containing the information is readily available to the employee. Thus the current Agricultural Wages Order can be posted up in a farm building or estate office. A booklet on sick pay schemes and pensions could be made available in a similar way.

Written information must be kept up to date, whether it is set out in a statement of terms or made available in other ways.

If an employer fails to provide an employee with a written statement, or it is incomplete, the employee can complain to an Employment Tribunal. Either party can go to a tribunal if the statement's accuracy is in doubt. The tribunal may determine the details of any particulars which are missing or amend those which are incorrect.

15.4.5 Implied terms

If there are any omissions in the agreed terms of the contract the court has the power to imply terms to make the contract workable. The court may also imply terms which are customary in a particular industry. In addition the courts imply certain terms unless the contract states otherwise. These include

(1) *The duty of mutual cooperation.* This means the employee must obey the employer's lawful and reasonable orders concerning the performance of his job. The employer must maintain a relationship of trust and confidence in his employee. For instance, he must investigate legitimate grievances. He must not insult his employee gratuitously nor berate him in front of his subordinates.

(2) *The employee's duty of faithful service.* Employees must serve their employers faithfully and honestly. They must not steal their employer's property nor make a secret profit from their job. They must not disclose trade secrets or confidential information.

   There is no general implied limitation on employees doing work in their spare time provided they do not damage their employer's business by, for example, working for a competitor. An employer who wants to restrict 'moonlighting' must make it a term of the contract.

(3) *The duty to take reasonable care.* An employer must take reasonable care for the safety of employees acting in the course of their employment. The Unfair Contract Terms Act 1977 prevents an employer from excluding this liability. Any contractual term which

restricts liability for death or personal injury from negligence is void. An employee must exercise reasonable skill and care in the exercise of his duty. Where an employer is liable vicariously to a third party for the negligent act of his employee, he is entitled to be indemnified by the employee. As it is unlikely that an employee will have adequate funds, employers are generally insured against most forms of liability. It is compulsory for employers to take out insurance against injury to their employees (see section 10.2).

15.4.6 *Changes to the contract*

Where contracts are made with reference to the Agricultural Wages Order the changes will automatically apply to the employee's contract. Changes may also be agreed expressly between the parties or by implication where the employee works on new terms. Unacceptable changes in a contract may result in a worker leaving. In this situation the employee may succeed in a claim that he has been constructively dismissed (see section 15.8). Or an employer may decide to terminate an employee's contract, after giving the appropriate notice, and offer a new contract to the employee. If the employee refuses the new contract he may be able to claim he has in fact been dismissed.

Changes in working conditions present problems. If a farm worker is asked to use new and more complicated machinery to do the same job, does this involve a contractual change? Generally speaking the answer is it will not. Workers are expected to keep up with modern technology. They are not, however, required to learn 'esoteric' skills.

15.5 Breaches of contract

An action for damages may be brought in the civil courts or the Employment Tribunal depending on the circumstances. Damages are assessed to compensate the injured party for losses which would normally arise from such a breach together with any losses which were reasonably foreseeable by the parties as a consequence of the particular breach.

Where there is a serious breach of contract the injured party can treat the contract as ended. An employee who, for instance, stole from his employer can be dismissed without notice. An employee who leaves as a result of a serious breach of contract by his employer may claim constructive dismissal.

15.6 Rights under the employment legislation

15.6.1 *General principles*

As stated above the contract is the starting point for ascertaining the terms of the employment. However, regardless of the contract, employment legislation gives the employee certain basic rights. Recent legislation has added to workers' rights the national minimum wage, entitlement to rest breaks and paid leave under the Working Time Regulations, parental leave and time off for domestic emergencies.

15.6.2 Pay

Since 1 April 1999 all relevant workers became entitled to the national minimum wage in accordance with the National Minimum Wage Act 1998. The legislation provides for a minimum hourly rate of pay, from which very little can be deducted. In particular there is a restriction on the amount that may be taken into consideration where living accommodation is provided by the employer.

Section 8 of the Employment Rights Act 1996 provides that all employees have a right to a written statement each time they are paid. This must show (1) the gross amount of the wages; (2) variable and fixed deductions; and (3) the net amount of wages. An employee can complain to an Employment Tribunal if an employer fails to provide a statement. The tribunal can order the employer to pay the employee a sum not exceeding the unnotified deductions made in the 13 weeks before the complaint was made. Where an employer makes a deduction from wages he must comply with section 13 of the Employment Rights Act 1996. He must ensure that he has the legal right to make the deduction. Deductions may be permitted by the contract (for example for misconduct, carelessness or other disciplinary reasons) or may be required by statute (for example tax and National Insurance contributions). Other deductions may be permitted on general legal principles (for example adjustments for small overpayments). Generally where a deduction is permitted by statute, written notification of the authorising term must be given to the employee before any deduction is made. Otherwise the worker must have agreed in writing to the deduction.

Most employees will have a right to sick pay under their contracts of employment. Notwithstanding any contractual term, or lack of terms, employers have a statutory obligation to pay Statutory Sick Pay (SSP) to all employees who pay Class I National Insurance contributions. There are three different rates, depending on the employee's normal weekly earnings. Employers can reclaim a proportion of the SSP from National Insurance contributions. Guidance leaflets are available from the DSS on the complex provisions relating to sick pay.

15.6.3 Right to notice

Most contracts will spell out the length of notice which an employer and an employee must give to terminate the employment. Section 86 of the Employment Rights Act 1996 lays down a *minimum* period of notice which the employer must give. He must give at least one week's notice to a person who has been continuously employed for more than four weeks but less than two years. Above two years the minimum period is one week's notice for every year of service up to a maximum of twelve weeks' notice. An employee who has been employed for more than four weeks must give one week's notice. These provisions do not prevent either side from waiving the right to notice or accepting payment in lieu. Nor do they affect the contractual period of notice when that stipulates a longer period.

15.7 Discipline

Every employer should have disciplinary rules and procedures for deciding when the rules have been broken. This means that both parties will know their position. It will be of use to an employer in any subsequent proceedings for unfair dismissal. The rules should be written into the statement of employment (see sections 15.4.3 and 15.4.4).

ACAS has issued a Code of Practice on Disciplinary Practice and Procedures in Employment. The Code states that employees should not be dismissed for a first disciplinary offence unless there is gross misconduct. The rules should state what conduct would merit immediate dismissal. Where an offence is less serious, the employee should be given a formal warning. It should be made clear that this is the first stage of the disciplinary procedure. A final written warning should be given if there is any further misconduct. This should make clear the consequences of any further breach of discipline.

15.8 Dismissal

15.8.1 On notice
The employer can dismiss an employee in accordance with the terms of the contract. He must make sure that he gives the length of notice required by the contract or by the employment legislation, whichever is the longer. If the proper notice is not given there will be a breach of contract and the employee can bring a claim in the Employment Tribunal.

15.8.2 Without notice
If an employee commits a very serious breach of contract the employer may dismiss him without notice. What constitutes a sufficiently serious breach depends on all the circumstances. Employers should ensure, where possible, that they have followed the disciplinary procedure described above.

15.8.3 Unfair dismissal
Section 94 of the Employment Rights Act 1996 gives employees a right not to be dismissed unfairly by their employer. It is possible to dismiss an employee for good reasons.

Certain employees are excluded from the protection of the Act. They

Employment

include those with less than one year's continuous employment on the date the contract ends and those who have reached the normal retiring age for the job. The normal retiring age is the age at which employees must retire unless special provision is made. If there is no such normal retiring age then it is taken to be 65. Husbands or wives of employers are also excluded from the provisions of the Act.

15.8.4 What is dismissal?

For the purposes of the legislation an employee is deemed to be dismissed in the following circumstances.

(1) Where the employer terminates the contract with or without notice. (If the termination is without notice, the employee may have a claim for breach of contract.) If an employee is given the option of resigning or being dismissed this counts as dismissal. Employees under notice who themselves give notice to leave earlier may still complain of unfair dismissal.

(2) Where the employee is employed under a fixed term which expires without being renewed.

(3) Where the employer's conduct entitles the employee to terminate his contract, with or without notice. This is known as constructive dismissal. Examples include failing to pay an employee's wages, changing hours or working conditions without agreement and breach of the duty of cooperation.

15.8.5 Fair reasons for dismissal

Under statute there are five reasons for dismissal which are said to be fair

(1) The employee lacks the capability or qualifications for the work for which he has been employed. Capability includes skill, aptitude, health or any other physical or mental quality.

(2) The employee's conduct.

(3) Redundancy (discussed in section 15.9 below).

(4) Where the continued employment would involve the employer or employee, or both, infringing legislation; for example, a delivery driver who has been disqualified from driving.

(5) Some other substantial reason which justifies the dismissal of an employee holding the position which the employee held. An example of these would be an employee's unreasonable refusal to agree to changes in the contractual terms. But employers would have to be able to show the tribunal good reasons for wanting to make the changes, such as reorganising to improve the efficiency of the farm or farm enterprise.

Employers must prove that they had a fair reason for dismissing the employee. That reason must have been present in the employer's mind at the time of the dismissal. It is always advisable to give written reasons for dismissal. A letter can be written to the employee explaining the reason. The worker is entitled to know why he is being dismissed whether for redundancy, inefficiency or sickness.

Employees with one year's continuous employment can demand a written statement from their former employers giving the reason for the dismissal. This must be supplied within 14 days. Where an employer unreasonably refuses to supply a statement, or makes an inaccurate or inadequate statement, the employee can complain to an Employment Tribunal. If the tribunal upholds the complaint, it must award the employee two weeks' pay.

15.8.6 Reasonableness

In addition to being a fair reason for dismissal it must be reasonable for an employer to dismiss the employee. The size and resources of the farm or farm enterprise will be a relevant factor. Where the dismissal is for misconduct the facts must support the claim. Moreover the employer should follow a fair procedure such as that set out in the ACAS Code of Practice. Where an employer wishes to change an employee's contractual terms as part of a business reorganisation, they should discuss the new terms with the employee and see if objections can be met. It will always be considered unfair to dismiss an employee for membership of an independent trade union or for participation in its activities.

The House of Lords considered the question of reasonableness in *Polkey* v. *AE Dayton Services Ltd* [1988] ICR 142. An employee had been made redundant without prior warning. An Industrial Tribunal (as the Employment Tribunal was then called) concluded that the result would have been the same whether or not he had been consulted. The House of Lords, however, ruled that the tribunal should have considered whether the employer had been reasonable or unreasonable at the time of the dismissal, and not whether the employee would have been dismissed in any event even if the proper procedures had been followed.

15.8.7 Remedies

An employee who has been unfairly dismissed has three remedies.

(1) *Reinstatement:* this requires the employer to treat the employee as if he or she has never been dismissed. Any arrears of pay or other benefits must be paid up.
(2) *Re-engagement:* the employee is employed in a new job which is comparable to the old or otherwise suitable.

    In deciding whether to order reinstatement or re-engagement the tribunal will consider the employee's conduct and wishes and whether it is practicable for the employer to comply with the order. These remedies are unlikely to be satisfactory where there is only a small farm workforce.
(3) *Compensation:* the basic award is calculated according to a formula.

An employee gets one and a half weeks' pay for each year of employment over the age of 41: one week's pay for each year of employment between the ages of 22 and 40, and half a week's pay for each year under the age of 22. The award is reduced by one twelfth for each month the employee is over 64. A maximum of 20 years service is taken into account. The maximum amount of a week's pay is currently £240. It is generally reviewed annually.

The award may be reduced where employees have caused or contributed to the dismissal, or their conduct before dismissal warrants a reduction. An unreasonable refusal of reinstatement will also result in a reduction. In addition the employee may be entitled to a compensatory award. The amount depends on what the tribunal considers 'just and equitable in the circumstances'. The maximum is £51 700.

15.9 Redundancy

15.9.1 Redundancy payments

Increased mechanisation and other labour-saving inventions have led to a reduction in the farm labour force. Workers who are made redundant are entitled to redundancy payments under employment laws. A farmer may find himself liable to make a redundancy payment to a farm worker, maintenance worker, gardener or domestic worker.

The following are not entitled to redundancy:

(1) Workers who have worked for an employer for less than two years
(2) Those who have reached the normal retiring age for the job
(3) Certain fixed-term employees
(4) Husbands and wives of employers
(5) Domestic servants who are close relatives of the employer
(6) Self-employed persons

The amount of redundancy payment is calculated in the same way as the basic award for dismissal except that service under the age of 18 does not count. The Department for Work and Pensions publishes a handbook on the Redundancy Payments Scheme, which includes a ready reckoner for calculating redundancy payments. A copy of the booklet can be obtained from the Department's offices.

In order to claim a redundancy payment the employee must have been dismissed. If employees have volunteered for dismissal, having been called upon to do so, they can still claim redundancy.

Redundancy arises in the following circumstances:

(1) Where the employer has ceased or intends to cease carrying on a business at the place where the employee was employed

(2) Where the work which the employees were employed to do has ceased or diminished or is likely to do so

Difficult questions arise where the nature or conditions of the work change. Employees are expected to adapt to new methods and technology. A farm secretary, for instance, would be expected to cope with the introduction of computers. On the other hand, where an employee is employed for a particular kind of work and that work ceases, then his dismissal will entitle him to claim a redundancy payment. So a farmer who sells his dairy herd may make his herdsman redundant. If he then employs a tractor driver to work the additional arable land, that will not affect the redundancy of the dairyman. But there will be no redundancy should the farmer dismiss one tractor driver and employ another in his place; nor if he dismisses an old man for unfitness and employs a younger man. It is a diminished requirement for the actual work not for the individual employee which gives rise to redundancy.

If an employee who has been made redundant is offered his old job back by the employer on the same terms as before, and he accepts that offer, the employee will not be entitled to a redundancy payment. An offer can be made orally or in writing, but it is advisable to make an offer in writing to avoid uncertainty and argument. An offer must be made before the notification of dismissal expires, and the new contract must start not later than four weeks after the end of the old contract. If the employee refuses the offer he will not get a redundancy payment unless he has reasonable grounds for his refusal.

If an employee is offered suitable alternative work by the same employer on different terms or in a different place and he accepts that offer, he will not be entitled to a redundancy payment. The offer can be made orally or in writing, but again it is advisable to make the offer in writing. It must be made before the notice of dismissal expires and the new contract must start not later than four weeks after the end of the old contract. At one time employees were reluctant to try a new job in case they lost their right to claim a redundancy payment as a result of leaving the new job because they found it unsuitable. In 1975 trial periods were introduced to avoid penalising workers willing to try a new job. The employee is allowed a trial period of four weeks in the new job to see whether or not he likes it. If he finds it is not suitable he can give notice and will be treated as though he had been dismissed on the date when the old job ended. However, if a worker on trial gives notice unreasonably he will not get his redundancy payment. If he gives notice on reasonable grounds he will get a redundancy payment.

A worker who unreasonably rejects the offer of a new job without giving it a trial will not get a redundancy payment. If he does accept the new job and continues working at it, he will not get a redundancy payment but his old job will count with his new job as continuous service so that if later he is made redundant he will have a payment based on employment in both jobs.

15.9.2 Death of an employer

If a farm business ceases as a result of the death of an employer, then his personal representatives will be liable to make redundancy payments to the employees of the deceased employer. However, in practice the personal representatives may carry on the farm business, at least for a time, and if they do so and they renew the contracts of employment or re-engage the employees under new contracts and the renewal or re-engagement takes effect within eight weeks of the death of the employer, a redundancy payment will not then be payable. If, later on, the workers are made redundant, then the personal representatives are liable to make a redundancy payment.

15.9.3 Employees who leave before their notice expires

An employee who has been made redundant may wish to leave before his period of notice expires. This is understandable, as he may want to take a new job which has been offered to him. Such a worker may not wish to be penalised by loss of his redundancy payment. He need not lose it. He should give his employer the statutory notice of one week, or a longer notice if this is required under the contract. If the employer does not object to the worker leaving he will still be entitled to his redundancy payment.

If the employer objects, he should serve notice on the employee requiring him to withdraw his notice and also warning him that if he fails to do so, his claim for redundancy payment will be contested. The employee might decide to ignore this request and leave just the same. If the employer then refuses to make a redundancy payment, the employee has a right to apply to an Employment Tribunal to decide whether or not he is entitled to such a payment. In order to take advantage of these provisions the employee's notice to the employer must be given within the period of notice the employer is obliged to give him, that is, under statute or contract not within a period of notice voluntarily given *(Lobb* v. *Bright Son & Company (Clerkenwell) Ltd* [1966] ITR 566).

15.9.4 Time off to look for work

An employee who has been made redundant and who has worked continuously for the same employer for at least two years, is entitled to reasonable time off during his period of notice to look for another job, or arrange for training for another job.

15.9.5 Unfair selection for redundancy

Where economies are being made in a farm or estate, more than one employee may have to be made redundant. The choice may not be easy. In businesses employing a large workforce, there is probably a 'first in last out' rule, or there may be a customary arrangement to that effect. On farms it is probably better not to have any rule. Such a rule can easily be applied, for example, on the railways, but where a farm employs one dairyman, one shepherd, one tractor driver, and two general farm workers, the rule would be very difficult to apply. Any proposed redundancy should be discussed with the employee and the situation explained.

15.9.6 Consultations with trade unions about proposed redundancies

Employers proposing to make workers redundant are required to consult recognised trade unions about redundancies. The union must be one which is recognised by the employers for the purposes of collective bargaining. The fact that a farmer pays the agricultural wage laid down in the Agricultural Wages Order does not amount to recognition by an employer of those unions involved in the fixing of the statutory minimum wage. The farmer pays that wage because the statutory order forbids him to pay less. He has not as an employer bargained with any union in respect of his employees. It seems that few farmers recognise a trade union for collective bargaining purposes.

Farmers who do recognise a trade union for collective bargaining purposes should consult the union in question about any proposed redundancies, even if it only concerns one employee. Any employer proposing to make 20 or more employees redundant is required to consult recognised trade unions within 30 days and in the case of 90 or more employees within 90 days. If an employer fails to consult a recognised trade union, that union can make a complaint to an Employment Tribunal, and the tribunal may make a protective award, which means that the employer will have to pay wages for a protected period, for example 90 days where 100 or more employees have been made redundant, 30 days where 30 or more have been made redundant, or 28 days in any other case. Complaints about failure to consult must be made within three months of dismissal. From 6 June 2000, the Employment Relations Act 1999 came into force, introducing a statutory recognition procedure to provide for a right to trade union recognition where a majority of the workforce wants it.

15.9.7 Notice of redundancies to the Secretary of State

In addition to consulting recognised trade unions an employer must notify the Secretary of State about redundancies.

Advance written notice of redundancies of more than 20 employees to be dismissed within 30 days must be notified to the Secretary of State in writing. Notice should be given 30 days before the first dismissal takes effect. For dismissals of 90 or more employees within 90 days, advance notice of 90 days is required; however, this provision is unlikely to be of interest to farmers.

15.9.8 Insolvent employer: redundancy payments

Where an employer cannot make a redundancy payment because he is insolvent, then subject to satisfactory proof being given to the Department of Work and Pensions, redundancy payments will be made from the Redundancy Payments Fund. The Department will then try to recover the money from the employers as an ordinary unsecured creditor. The Department of Employment advises that an employer who admits liability for redundancy payments, but is unable to pay them, should explain the problem to the nearest office of the Employment Services Agency and be ready to send a statement of account or a written statement from an accountant or solicitor setting out his financial position. The Department also advises that legal representatives of deceased employers who expect delays in the settlement of the estate should explain the position to the nearest employment office for reference to the specialist office of the Department.

15.9.9 Time limit for claims

As a rule, redundancy payments will be made at the time of the dismissal, but an employer may fail to make a payment either because he is insolvent or for some other reason. An employee has six months from the date of termination of employment in which to make a claim which should be made in writing. After the expiry of the six-month period any liability by the employer to make a payment will cease, if no claim has been made in writing.

If a claim in writing has been made within the six-month period, but refused, the employee has a further six months in which to refer the claim to the Employment Tribunal.

15.10 Change in ownership of business

The Transfer or Undertakings (Protection of Employment) Regulations 1981 give employees certain rights if an undertaking, or part of it, is transferred from one employer to another. The Regulations apply where the transfer is made by way of a sale or by some other disposition or by operation of law, but only apply to commercial undertakings.

Where there is a transfer of a commercial undertaking, the transfer will not terminate the employee's contract of employment. Persons still employed by the business immediately before the transfer will not have their contracts of employment terminated by the transfer. Instead, any contracts of employment which would otherwise have been terminated by the transfer will continue as though they had been made between the employees and the new owner of the business.

It has been held by the High Court that where a farm changes hands because there is a grant of a new tenancy to a new tenant, or a tenanted farm is taken in hand, or an owner-occupied farm is sold, there is a change in the ownership of a farm business (*Lloyd* v. *Brassey* [1969] 1 All ER 382). It appears therefore that the Transfer of Undertakings (Protection of Employment) Regulations 1981 will apply in all those situations where employees are still employed by the outgoer immediately before the transfer. Generally a farmer will want to continue to employ workers up to the time of transfer so as to hand over the farm in good order. However, there will be cases where workers will be made redundant before that date. For example, the outgoer may decide to sell his herd or flocks some months before he has contracted to sell the farm. He could then make his herdsman or shepherd redundant and make the statutory payments for redundancy.

An employee who is dismissed because of a transfer of a business is to be treated as unfairly dismissed, that is if the transfer of the business is the principal reason for his dismissal. This provision could lead to considerable difficulties. For example, if an owner of a farm sells his herd some months before he contracts to sell the farm, and makes his herdsman redundant, that would not be a dismissal for reasons of the transfer. If, however, he sells the herd after he has contracted to sell the farm but before the conveyance is completed, then the dismissal would seem to be a dismissal by reason of transfer. However, the rule is modified by provisions in regulation 8(2) which states that dismissal shall not be treated as unfair if it is for an economic, technical or organisational reason entailing changes in the workforce of either the outgoer or incomer before or after the transfer. Thus the dismissal of a herdsman after a contract of sale, but some months before the completion of the conveyance of the farm, might be considered to be an economic or organisational reason for dismissal.

As far as the incomer is concerned, if he dismisses the employees who have been transferred to him as a result of the change in the ownership of the business, he is entitled to dismiss them for economic, technical or organisational reasons, without this being treated as automatically unfair. However, he would still be liable to make redundancy payments which would be calculated on the basis of continuous employment with the business transferred to him.

The general purpose of the Regulations seems to be that workers shall not be dismissed simply because a business is transferred, but that if they are dismissed for economic, technical or organisational reasons, they will get their redundancy payments.

This proposition is illustrated by the case of *Litster* v. *Forth Dry Dock Co. Ltd* [1989] 1 All ER 1134. The receivers of an insolvent company (who were not in a position to pay any redundancy money, any sum in lieu of notice or any accrued holiday pay) dismissed the employees an hour before the undertaking was transferred to a new company. The House of Lords held that the new owners were liable to pay compensation for unfair dismissal. There was no evidence to show the employees had been dismissed for 'economic, technical or organisation reasons entailing changes in the workforce'. The dismissal was unfair because it was for a reason connected with the transfer. The new owners had hoped thereby to avoid the liability for unfair dismissal or redundancy claims which would arise if the dismissals were made after the transfer.

15.11 Health and safety at work

Under the 1974 Health and Safety at Work Act employers have a duty to ensure so far as is 'reasonably practicable' the health, safety and welfare at work of all employees. This involves the employer in maintaining safe premises, a safe system of work and the provision of information, training and supervision. An employer must also comply with the Noise at Work Regulations 1989 (SI 1989/1790) amongst a variety of regulations that may apply in particular the Control of Substances Hazardous to Health Regulations 1999 (SI 1999/437) and the Reporting of Injuries, Diseases and Dangerous Occurrences Regulations 1995 (SI 1995/3163).

Employers of five or more employees must have a written statement of their health and safety policy. They must bring it to the attention of their employees.

Employers also have a duty to run their businesses to ensure as far as reasonably practicable that non-employees are not exposed to any risk to their health and safety. For instance they should not allow dangerous animals to roam on public footpaths. Employees too must take reasonable care for themselves and their fellow workers who may be affected by their acts and omissions.

Health and Safety Inspectors have powers to enter premises and carry out investigations. They can instigate prosecutions for breaches of the law in the magistrates' court. Or they can serve notices on employers requiring a person to act or refrain from acting in the matters specified in the notice. A person served with a notice can appeal to the Employment Tribunal.

Chapter 16
Rating and Council Tax

Abbreviation in this chapter
'1988 Act'   Local Government Finance Act 1988
'1992 Act'   Local Government Finance Act 1992
'1997 Act'   Local Government and Rating Act 1997
'2001 Act'   Rating (Former Agricultural Premises and Rural Shops) Act 2001

16.1   Rating

16.1   Non-domestic rates
The liability of non-domestic property to rates is governed by the Local Government Finance Act 1988, although this re-enacts much of the General Rate Act 1967 and therefore case law in relation thereto continues to be of some relevance.

16.1.2   Rating of domestic property
Since domestic rating was abolished in 1990, much may turn on whether a property is domestic or not. Section 66 of the 1988 Act makes various provisions in relation to this. It states, 'property is domestic if it is used wholly for the purposes of living accommodation' and the homestead's garden, yard, garage and other structures are included. Special provision is made for bed and breakfast (see section 16.2.2 below), and self-catering accommodation available for letting for short periods of time totalling 140 days or more is adjudged to be non-domestic.

16.1.3   The rateable value
Rates are payable at the rate of 41.6 pence in the pound (41.2 pence in Wales) on the rateable value. Paragraph 2 of Schedule 6 to the 1988 Act defines the rateable value as follows:

> 'The rateable value of a non-domestic hereditament shall be taken to be an amount equal to the rent at which it is estimated the hereditament might reasonably be expected to let from year to year if the tenant undertook to pay all usual tenant's rates and taxes and to bear the cost of the repairs and insurance and the other expenses (if any) necessary to maintain the hereditament in a state to command that rent.'

This hypothetical rent payable by a hypothetical tenant may not be the same as the actual rent paid by the actual tenant where the premises are actually let, although this may be evidence of what the hypothetical tenant might pay.

16.1.4 Rating list

It is the responsibility of the local valuation officer of the Inland Revenue to enter property in the local rating list. There is no obligation on a ratepayer to initiate an entry in the list in respect of an activity which is liable to rates. However, where an entry is made in the current rating list, liability to rates can be backdated to the date the activity commenced. A liability for back rates on an activity will only be avoided after the first anniversary of the date on which the next rating list is compiled.

The present rating list dates from 1 April 2000 and is based on rental values at 1 April 1998. The next rating list will commence on 1 April 2005.

When a new entry is made in the list the valuation officer has to send the ratepayer a notice giving details of the entry. The ratepayer has six months from the date of that notice in which to appeal; such appeals must be in writing and contain certain prescribed information. Appeal forms, called *Proposal to Alter the Rating List*, are obtainable from valuation offices. If an appeal is not resolved by negotiation, it is referred to a valuation tribunal and a hearing is held. Whilst an appeal is outstanding the ratepayer is liable to pay the rates as assessed. Where that liability is reduced, any overpayment will be refunded with interest.

16.1.5 Special rural rate reliefs

The Rating (Former Agricultural Premises and Rural Shops) Act 2001 introduced a special new relief for farm diversifications. The key points of this relief are as follows:

- The Act provides for a 50% mandatory rate relief for land and buildings used for non-agricultural purposes on what has been agricultural land or agricultural buildings for at least 183 days during the year prior to the date in which the provisions come into force (15 August 2001 in England).
- Local authorities have a discretionary power to increase the relief to up to 100% where they feel that the change of use will be of benefit to the rural community and that the cost to the council taxpayer is justified.
- Both the mandatory and discretionary relief are initially limited to a maximum of five years from the date of commencement of the Act, with a provision for this to be extended by order. However, premises that had already benefited from the relief before any extension will only receive relief for a maximum of five years from the date the original premises first qualified for the relief.
- This rate relief will only apply to hereditaments with a rateable value less than the prescribed amount, which is currently £6000 in England under the Non-Domestic Rating (Former Agricultural Premises) (England) Order 2001 (SI 2001/2585).

In addition, under the 1997 Act, local authorities have had the discretion to grant up to 100% relief for businesses in rural areas with a rateable value of less than £12 000. However, in exercising this discretion, the local authority has to be satisfied that that the business is of benefit to the rural community and the cost to the council taxpayer is justified (the council bears 25% of the cost of such reliefs). Therefore a special case would have to be shown, for instance because the activity was targeted at children or the people with disabilities, or created local employment.

Under the 1997 Act, there is also mandatory relief for sole post offices and general stores in rural areas with a rateable value of £6000 or less, with a discretion to grant relief for as much again. The 2001 Act extends this to all village shops that sell mainly food for human consumption.

16.1.6 Unoccupied property

Unoccupied non-domestic hereditaments on a rating list are rateable with a 50% reduction, but the Secretary of State can make regulations prescribing exemptions (sections 45 and 46 of the 1988 Act). The exempt categories are to be found in the Non-Domestic Rating (Unoccupied Property) Regulations 1989) SI 1989/2261, as amended by SI 1995/549). These include

- The first three months of non-occupation
- Properties with a rateable value of less than £1900
- Certain industrial premises, namely those used for manufacturing, repairing or adapting goods or materials, storage, the working or processing of minerals, and the generation of electricity, provided none of the buildings are for the retail provision of goods or services
- Where the owner is prohibited by law from occupying the hereditament
- Where the building is listed and unoccupied
- Where the hereditament is included in the Schedule of Monuments compiled under section 1 of the Ancient Monuments and Archaeological Areas Act 1979

There is also a notice procedure for establishing when unoccupied newly erected buildings are to be deemed to be completed for rating purposes: (see section 46A of the 1988 Act), so as to prevent deliberate delays in completion in order to avoid attracting rating liability.

16.1.7 Charities

There is a mandatory 80% rating relief for hereditaments of charities wholly or mainly used for charitable purposes. Charging authorities have a discretion to increase the relief as far as 100% (sections 47 and 48 of the 1988 Act).

16.1.8 Discretionary relief for recreational clubs etc.

In addition, sections 47 and 48 of the 1988 Act give local authorities a discretion to reduce or eliminate the rates payable by non-profit making organisations including recreational, philanthropic, religious, educational, scientific, welfare, literary and artistic organisations.

16.1.9 Agricultural land

Land is a rateable hereditament (section 64(4) of the 1988 Act), but land and buildings used *solely* for agricultural purposes are exempt from rating (Schedule 5).

'Agricultural land' is defined in paragraph 2(1) of Schedule 5 as

'(a) Land used as arable, meadow or pasture ground only,
 (b) Land used for a plantation or a wood or for the growth of saleable underwood,
 (c) Land exceeding 0.10 hectare and used for the purposes of poultry farming,
 (d) Anything which consists of a market garden, nursery ground, orchard or allotment (which here includes an allotment garden within the meaning of the Allotments Act 1922), or
 (e) Land occupied with, and used solely in connection with the use of, a building which (or buildings each of which) is an agricultural building.'

Paragraph 2(2) of Schedule 5 goes on to say agricultural land does not include 'land occupied together with a house as a park, gardens (other than market gardens), pleasure grounds, land used *mainly or exclusively* [emphasis added; see section 16.1.10] for the purposes of sport or recreation, or land used as a racecourse' (Schedule 5, para 2 (2)).

Section 64(11) of the 1988 Act provides that 'land' includes 'land covered by water'. Any land covered by water which is predominantly agricultural will not be rateable (see *Watkins* v. *Hereford Assessment Committee* (1935) 154 LT 262; and *Garnetts* v. *Wand* (1960) 7 RRC 99).

16.1.10 Mixed uses

Where land normally used as arable, meadow or pasture ground is occasionally used for a non-agricultural purpose (e.g. gymkhana or Pop Festival) it should be noted that the word 'only' occurs in para (a) of the definition quoted at section 16.1.9 above. This stricture would not seem to be qualified by schedule paragraph 2(2), although it will still be subject to the *de minimus* rule. This is illustrated in *Hayes* v. *Loyd* [1985] 2 All ER 313, the House of Lords held that some land used for grazing except for one point-to-point meeting a year was rateable as 'land used for a racecourse', because it was laid out with permanent rails, jumps and other sophisticated racecourse facilities and the annual event was big and commercially profitable. On the facts of the case the racing was not to be ignored as *de minimis*. The court explained that in different circumstances the racing activity might be so little or informal as not to make the land rateable.

16.1.11 Agricultural buildings

The Act defines agricultural buildings exempt from rating. By paragraph 3 of Schedule 5

> 'A building is an agricultural building if it is not a dwelling and:
>
> (a) it is occupied together with agricultural land and is used solely in connection with agricultural operations on the land, or
> (b) it is or forms part of a market garden, and is used solely in connection with agricultural operations at the market garden.'

In interpreting the word 'solely' no account shall be taken of any time during which a building 'is used in any other way, if that time does not amount to a substantial part of the time during which the building is used' (paragraph 8(3) Schedule 5)).

A case in Yorkshire in 2000 (unreported) where a farmer who also undertook contract farming was held not to be eligible for the agricultural exemption in respect of the buildings in which his machinery was stored on the basis that the buildings were not used 'solely' in connection with agricultural operations on the land with which they were occupied, as is required by the rating legislation. Following this a consultation paper was issued proposing an extension of the definition of 'agriculture' in the rating legislation so that it covers machinery rings and share/contract farming. At the time of writing (2002) these proposals have not been taken forward.

16.1.12 Syndicates' buildings

The 1988 Act provides that buildings of farming syndicates are excluded from rating, although they might be occupied by a management committee for separate occupiers of agricultural land. The provision (paragraph 4 of Schedule 5) is a little complex, but essentially as long as the building is occupied by all the occupiers of the land it serves, or by persons appointed by them each of whom occupies some of the land, the agricultural building gets rating exemption, provided there are not more than 24 occupiers of the land. In Schedule 5, paragraph 7 ensures that the exemption extends to incorporated associations and cooperatives, provided the building is not used in connection with land occupied by a non-member of the body. A marketing society's auction hall was held not exempted from rating where a substantial quantity of the goods sold belonged to non-members (*Corser* v. *Gloucestershire Marketing Society Ltd* (1981) 257 EG 825 (CA)).

16.1.13 Livestock buildings

Even if not 'used solely in connection with agricultural operations' on the agricultural land as stated in the definition of 'agricultural building', a building used for keeping or feeding livestock is an agricultural building exempt from rating provided

(1) it is solely used for that purpose; or
(2) it is occupied with agricultural land and its sole use is in connection with agricultural operations on that land and keeping or breeding livestock; and
(3) it is surrounded by or contiguous to an area of agricultural land which amounts to not less than two hectares.

This is a slight simplification of the rather complex provisions to be found in paragraph 5 of Schedule 5 to the 1988 Act.

'Livestock' in the 1988 Act is stated to 'include' any mammal or bird kept for the production of food or wool or for the purpose of its use in the farming of land. Rating exemption does not therefore extend to buildings for keeping animals in wildlife parks or zoos.

Game birds not raised for food but for release are not 'livestock' (*Cook v. Ross Poultry Ltd* [1982] RA 187).

As bees are unlikely to be interpreted as 'livestock' the 1988 Act makes special provision to exempt from rating any building (other than a dwelling) occupied by a person keeping bees, if it is used solely for that purpose and it is surrounded by or adjoins at least two hectares of agricultural land (paragraph 6 Schedule 5).

The House of Lords, in *Hemens (VO)* v. *Whitsbury Farm & Stud Ltd* [1988] 1 All ER 72, held that buildings for breeding and keeping riding horses were not exempt from rating, but the land on which the horses grazed was exempt agricultural land. Following this decision the government was persuaded to give a measure of exemption of buildings for rearing and keeping horses and ponies, where it was a minor activity on farms. The first £3000 of rateable value is exempt from rating assessment (The Non-Domestic Rating (Stud Farms) (England) Order 2001, SI 2001/2586).

16.1.14 *Rates on sporting rights*
Rates on sporting rights, where they were enjoyed separately from the land over which they existed, were abolished by section 2 of the 1997 Act with effect from 1 April 1997.

16.1.15 *Fish farms and fisheries*
Land and buildings used solely for, or in connection with, aquaculture (including shellfish farming) are exempt from rating, whether the fish are reared directly for food or for stocking fisheries (1988 Act, paragraph 9 Schedule 5).

In some instances fisheries are rated as land. Section 64(11) of the 1988 Act states what is included in 'land', and, although it is not a comprehensive definition, it is significant that it is not stated (as in other statutes) to include 'land covered by water' (there was no definition of 'land' when

Thomas v. *Witney Aquatic Co. Ltd* (1972) RRC 348 was decided by the Lands Tribunal). Any land covered by water which is predominantly agricultural will in any case not be rateable (see *Watkins* v. *Hereford Assessment Committee* (1935) 154 LT 262; and *Garnetts* v. *Wand* (1960) 7 RRC 99).

16.1.16 Some useful cases

The cases listed here provide further insight into this complex area.

- *Handley* v. *Bernard Matthews plc* [1988] EGCS 129 (LT): a mill producing pellets for turkeys on 29 farms owned by the same company between 9 and 74 miles from the mill was held to be an exempt agricultural building 'occupied together with' the farms because the Act did not impose a geographical test.

- *Fitter* v. *Fraser-Smith* [1988] RA 231 (LT): a grass drying plant making grass pellets for feed, with office premises was held exempt from rating, where the grass was grown by four operators and sold to the company under a scheme of agreements.

- *Fletcher* v. *Bartle* [1988] RA 284 (LT): a farm shop selling only produce from the farm was held rateable because retail sales were not the same purpose as the activities carried on on the associated agricultural land.

- *Courtman (VO)* v. *West Devon & North Cornwall Farmers Ltd* [1990] RA 17: a feedmill used by a farming cooperative so that 98% of its output was used on members' farms was exempt from entry on the valuation list as an agricultural building.

- *Farmer (VO)* v. *Hambleton DC* [1993] 1 All ER 117: a ratepayer owned a poultry processing factory, along with many poultry farms. The factory was some distance away from the farms. It was held that for a building to be an agricultural unit it had to be occupied together with agricultural land in a physically contemporaneous unit, and this was not the case here.

- *Womersley (Valuation Officer)* v. *Jisco* [1990] RA 211: buildings used for the slaughter of animals and dressing of the carcasses, where the livestock was raised on the farm on which the buildings stood, were agricultural buildings that were exempt from rating.

- *Covell (Valuation Officer)* v. *Littman* (1984) 272 EG 797: the premises used for cheesemaking; the cheese was made entirely from milk produced on the premises, and almost all the milk produced there was used for the cheese. It was held that the cheesemaking was a single and coherent operation in agricultural production, consequent upon the agricultural operation of grazing cattle on land, whilst the end result was clearly recognisable as agricultural produce. Consequently the cheesemaking was a use in connection with operations on agricultural land, and the exemption was allowed.

16.2 Council tax

16.2.1 Background
The system of local taxation on domestic property changed radically in the early 1990s. The long standing system of domestic rates was replaced by the short-lived community charge (or poll tax as it was commonly known) on 1 April 1990. The unpopularity of this tax led to its abandonment after three years, and the property-based council tax was introduced by the Local Government Finance Act 1992, commencing on 1 April 1993.

16.2.2 On what property is council tax payable?
Council tax is payable in respect of any dwelling, unless it is exempt. Composite hereditaments are also dwellings for the purposes of council tax. A property or part thereof used wholly or mainly in the course of a business for short-stay accommodation falls to be rated as a non-domestic hereditament. However, by special provision under section 66(2A) of the 1992 Act a property is subject to council tax where

- Short-stay accommodation is provided for less than 6 people
- The provider has his sole or main residence within the hereditament
- The short-stay accommodation is subsidiary to the use of the hereditament as a sole or main residence

Where part of a house is used as an office, this will turn the premises into a composite hereditament, and the extent of the property liable to rates will turn on the degree of business user (see *Fotheringham* v. *Wood (VO)* [1995] RA 315.

Certain descriptions of property cannot in themselves constitute dwellings for council tax, despite their domestic use, unless they form part of a larger property which is a dwelling, namely a yard, garden, outhouse, private garage, private storage used wholly or mainly for the storage of articles of domestic use. Therefore if any of these types of property are separate from a dwelling, they will be liable to rates.

By virtue of the Council Tax (Chargeable Dwellings) Order (SI 1992/549) where a building or part of a building is constructed or adapted for use as separate living accommodation, this will be treated as a separate dwelling for council tax purposes. Planning restrictions on the separate use of the units are not necessarily relevant to this test. Further the degree of communal living and whether the annexe is capable of being separately sold are irrelevant considerations: see *Rodd* v. *Ritchings* [1995] RA 299. Following this decision, a special exemption was introduced from 1 April 1997 specifically exempting from council tax self-contained accommodation with another dwelling when the accommodation is the sole or main residence of a person over the age of 65 who is the dependent relative of a person who is resident in the other dwelling.

Once the domestic status of the property has been established, there must be compelling evidence that its character has changed substantially if reclassification as a non-domestic hereditament is to be justified: see *Guthrie* v. *Highland Region Assessor* [1995] RA 292.

There are a number of exemptions from council tax, the most relevant of which to countryside property are listed below.

- *Class A:* A dwelling which has been vacant for less than 12 months and which requires or is undergoing major repair work to render it habitable or is undergoing structural repairs, or has been vacant for less than six months from the day on which such works were substantially completed.
- *Class C:* A dwelling which has been vacant for less than 6 months.
- *Class F:* A dwelling which is unoccupied following the death of the occupier where either probate has not been granted or less than six months have elapsed since its grant.
- *Class G:* Where occupation is prohibited by law (e.g. certificate of unfitness for human inhabitation).
- *Class R:* A pitch or mooring not occupied by a caravan or boat.
- *Class T:* An unoccupied dwelling which cannot be let separately from the rest of the property without a breach of planning control.
- *Class W:* Occupation of self-contained units by relatives over the age of 65 (see above).

16.2.3 Who is liable to pay the council tax?

The general rule is that liability depends upon residence of a dwelling. In order of priority, it falls as follows:

(1) A resident freeholder
(2) A resident leaseholder (including a tenant holding under an assured tenancy) whose interest is not inferior to another such interest held by another such resident
(3) A resident statutory or secure tenant
(4) A resident who has a contractual licence to occupy the whole or any part of the dwelling
(5) A resident
(6) The owner of the dwelling

Liability will only fall on one of the above categories of people. Where there is no person answering the description in the first category, liability falls to the person in the next category and so on. Where two or more persons fall into the applicable category, they are jointly and severally liable to pay the council tax in respect of the dwelling. Council tax is calculated on a daily basis.

The owner of a caravan or boat which constitutes a dwelling is liable to council tax except in respect of those days when a person other than the owner is resident and so becomes liable for those days.

16.2.4 Valuation and banding

Council tax is payable in accordance with eight valuation bands. The value of each dwelling is assessed by the valuation office agency on the following assumptions:

(1) The dwelling was sold on 1 April 1991 taking account of any change between then and (broadly) 1 April 1993
(2) Any sale with vacant possession, and in the case of a house which was sold freehold, in the case of a flat on a 99 year lease
(3) The dwelling was in reasonable repair
(4) The use of the dwelling was permanently restricted to use of a private dwelling, and the dwelling has no development value

The bands are as follows:

	England			Wales		
A	£0	–	£40 000	£0	–	£30 000
B	£40 000	–	£52 000	£30 001	–	£39 000
C	£52 001	–	£68 000	£39 001	–	£51 000
D	£68 001	–	£88 000	£51 001	–	£66 000
E	£88 001	–	£120 000	£66 001	–	£90 000
F	£120 001	–	£160 000	£90 001	–	£120 000
G	£160 001	–	£320 000	£120 001	–	£240 000
H	over £320 000			over £240 000		

Therefore changes in house prices generally since 1991 will not affect the banding of a property. However, there are a limited number of circumstances in which dwellings may be rebanded. These are listed below.

(1) Where there has been a material increase in value since the valuation band was first shown in the list. The alteration may only be made when the dwelling is sold, or has become subject to a lease of seven years or more. This might apply where an extension has been added to the property.
(2) Where there has been a material reduction in value caused by the demolition of any part of the dwelling or any change in the physical state of the dwelling's locality. Such events may give rise to an immediate rebanding.
(3) Where the dwelling has become or ceased to become a composite hereditament but there has been an increase or reduction in its domestic use, or where there is a successful appeal against the banding. These may only be done by a new owner, within six months of their assumption of ownership.

The valuation of composite hereditaments was considered by the Divisional Court of the Queen's Bench in the case of *Atkinson* v.

Cumbrian Valuation Tribunal, The Times 30 July 1996. The court held that it was not always necessary to make a valuation of the whole hereditament (in this case a whole farm including a farmhouse) in order to reach the capital value reasonably attributable to the domestic element only, and that this could be reached by establishing a separate value of the farm house only, and then making an appropriate adjustment (in this case 10%) for the fact that it formed part of a larger hereditament. This was underlined by the fact that council tax was levied by reference to valuation bands, and did not require specific values to be placed on the dwellings.

16.2.5 Discounts

There is a discount of 25% where the dwelling constitutes the sole or main residence of *one* person only. A discount of 50% is given where the dwelling does not constitute any person's sole or main residence (e.g. a week-end home). There are various other 'status' discounts, such as for prisoners and students, who are then disregarded in applying for the above discount purposes.

Section 12 of the 1992 Act allows the Secretary of State, in the case of Wales where there is a multiplicity of second homes, to prescribe classes of dwellings in respect of which the billing authority then has the power to determine that, whether or not there are residents of dwellings in those classes, the discounts should be 25% instead of 50% or that there should be no discount. This power is exercised through the Council Tax (Prescribed Classes of Dwellings) (Wales) Regulations 1998 (SI 1998/105). In November 2001 the government consulted on proposed changes to the reliefs on second homes and long term empty homes. At the time of writing (2002) these proposals have not been taken forward.

Further Reading

Chapter 1
Harpum C. (2000) *Megarry & Wade: the law of real property*. 6th edn. Sweet & Maxwell, London.

Chapter 2
Muir Watt & Moss (1998) *Agricultural Holdings*. 14th edn. Sweet & Maxwell, London.
Scammell, W.S. (1997) *Scammell & Densham's Law of Agricultural Holdings*. 8th edn. Butterworths Law, London.
Sydenham, A. & Mainwaring, N. (1995) *Farm Business Tenancies*. Jordans, Bristol.

Chapter 3
Muir Watt & Moss (1998) *Agricultural Holdings*. 14th edn. Sweet & Maxwell, London.
Scammell, W.S. (1997) *Scammell & Densham's Law of Agricultural Holdings*. 8th edn. Butterworths Law, London.

Chapter 4
Rodgers, C.P. (1998) Farm Cottages. In: *Agricultural Law*. 2nd edn. Butterworths, London.
Williams & Johnstone (2000) *Farm Cottages*. Burges Salmon, London.

Chapter 5
Furber, J. (1999) *Hill & Redman's Guide to Landlord and Tenant Law*. Vol. 1. Butterworths Law, London.
Lewison, K. (1978) *Woodfall: Landlord and Tenant*. Vol. 2. Sweet & Maxwell, London.

Chapter 6
Sydenham, A. (2001) *Public Rights of Way and Access to Land*. Jordans, Bristol.

Table of Cases

A-Gen v. Chambers (1854) 23 LJ Ch. 662 195
AG v. Squires (1906) 5 LGR 99 165
Alconbury Developments v. Secretary of State DETR [2001] EGCS 5
 96
Allen v. Greenwood [1979] 2 WLR 187 164
Allen v. Thompson (1870) LR 5 QB 336 189
Alphacell Ltd v. Woodward [1972] 2 All 475 HL 207
Anns v. Merton London Borough [1978] AC 728 162
Atkinson v. Cumbrian Valuation Tribunal, *The Times* 30 July 1996
 264

Baker v. Berkeley (1827) 3 C&P 32 200
Batchelor v. Kent County Council [1990] 1 EGLR 32 126
Birch v. Mills (Unreported) 80
Bland v. Yates (1914) 58 S.J. 612 163
Bolton v. Stone [1951] AC 850 164
Bone v. Seale [1975] 1 All ER 797 165
Bracey v. Read [1963] Ch 88 62
British Railways Board v. Herrington [1972] AC 877 161
British Waterways Board v. Anglian Water Authority, *The Times*
 23 April 1991 176
British Waterways Board v. Severn Trent Water Ltd (Judgement
 2 March 2001) 154
Browne v. Marquis of Sligo (1859) 10 Ch 1 195
Burditt v. Joslin [1981] 3 All ER 203 192
Burgess v. Gwynedd River Authority (1972) 24 P&CR 150
 182

Cambridge Water Co. Ltd v. Eastern Counties Leather plc [1994]
 2 AC 264 169
Cargill v. Gotts [1981] 1 All ER 682 179
Chasemore v. Richards (1859) HL Cas 349 176
Childers v. Anker [1995] EGCS 116 35
Collins v. Thames Water Utilities Ltd [1994] 49 EG 116 .. 155
Cook v. Ross Poultry Ltd [1982] RA 187 259
Corser v. Gloucestershire Marketing Society Ltd (1981) 257 EG 825
 (CA) 258
Cottle v. Coldicott [1995] SpC 40 230
Courtman (VO) v. West Devon & North Cornwall Farmers Ltd
 [1990] RA 17 260
Covell (VO) v. Littman (1984) 272 EG 797 260

Table of Cases 267

Cunliffe v. Bankes [1945] 1 All ER 459 166

Director of Land and Buildings v. Shun Fung Ironworks Ltd. [1995]
  2 WRL 404 132
Donovan v. Dwr Cymru [1994] 1 EGLR 203 156

Empress Car Co. Ltd v. NRA [1977] Env..LR 227 208
Essexcrest v. Even (1988) 55 P&CR 279

Farmer (VO) v. Hambleton DC [1993] 1 All E.R. 117 260
Faulks v. Faulke [1992] 15 EG 82 229
Ferae Naturae in London Borough of Wandsworth v. Railtrack plc
  165
Fitter v. Fraser-Smith [1988] RA 231 (LT) 260
Fitzhardinge (Lord) v. Purcell [1908] 2 Ch 139 187
Fletcher v. Bartle [1988] RA 248 (LT) 260
Fletcher v. Rylands (1866) LR 1 Ex 265 168
Fotheringham v. Kerr (1984) 48 P&CR 173 194
Fotheringham v. Wood (VO) [1995] RA 315 261

Garnetts v. Wand (1960) 7 RRC 99 .... 257, 260
Giles v. Walker (1890) 24 QBD 656 .... 164, 166
Goodman v. Saltash Corporation (1882) 7 App. Cas 633 195
Guthrie v. Highland Region Assessor [1995] RA 292 262

Hanbury v. Jenkins [1901] 2 Ch 401 194
Handley v. Bernard Matthews plc [1988] EGCS 129 (LT) 260
Hanning v. Top Deck Travel (1993) 68 P&CR 14 4
Harries v. Barclays Bank plc 203
Hayes v. Loyd [1985] 2 All ER 313 257
Hemens (VO) v. Whitsbury Farm & Stud Ltd [1988] 1 All ER 72
  .... 257, 259
Hertfordshire County Council v. Ozanne [1989] 2 EGLR 18 126
Hidson v. Ashby [1896] 2 Ch 1 194
Holbeck Hall Hotel v. Scarborough Borough Council [2000]
  2 All ER 705 CA 164
Holden v. White [1982] QB 679 159
Horn v. Sunderland Corporation [1941] 2 KB 26 131
Howkins v. Jardine [1951] 1 All ER 320 32
Hunter v. Canary Wharf Ltd [1997] A.C. 655 163

Inglewood Investment Co. v. Forestry Commission [1989] 1 All ER 1
  189

Jelley v. Backman [1974] QB 488 20
Johnson v. Moreton [1978] 3 All ER 37 36

Lathall v. Joyce & Sons [1939] 3 All ER 854 169
League Against Cruel Sports v. Scott [1985] 2 All ER 489 200
Leakey v. National Trust [1980] 1 All ER 17 .... 164, 165
Leeman v. Montagu [1936] 2 All ER 1677 165
Lemmon v. Webb [1896] AC 1 168
Litster v. Forth Dry Dock Co. Ltd [1989] 1 All ER 1134 252
Lloyd v. Brassey [1969] 1 All ER 328 252
Lobb v. Bright Son & Company (Clerkenwell) Ltd [1996] ITR 566
  249
Lovett v. Fairclough (1990) 61 P&CR 385 194

McKinnon Industries Co. Ltd v. Walker [1951] 3 DLR 577, 581
  164
Malone v. Laskey [1907] 2 KB 141 163
Marcic v. Thames Water Utilities Ltd [2001] 3 All ER 169
Marriage v. East Norfolk Rivers Catchment Board [1949]
  2 All ER 1021 182
Mason v. Clarke [1955] 1 All ER 914 185
Matthews v. Wicks, *The Times* 25 May 1987 (CA) 171
Mercury Communications Ltd v. London & India Dock Investments
  [1995] 69 P&CR 135 146
Metropolitan Board of Works v. McCarthy [1974] LR 7 HL 243
  133
Miller v. Jackson [1977] QB 966 164
Millington v. Secretary of State and Shrewsbury & Atcham BC [2000]
  JPL 297 94
Mills v. Avon and Dorset River Board [1953] 1 All ER 382 198

Neil v. Duke of Devonshire (1882) 8 App Cas. 135 195

Paget v. Birkbeck (1863) 3 F&F 683 200
Pannett v. P G Guinness & Co. Ltd [1972] 2 QB 599 161
Patel v. W H Smith (Eziot) Ltd, *The Times* 16 February 1987 ... 172
Pattinson v. Finningley Internal Drainiage Board (1971) 22 P&CR 929
  183
Paul v. Summerhayes [1874] 4 QBD 9 (DC) 200, 201
Peaty v. Field [1970] 2 All ER 895 165
Peech v. Best [1931] KB 1 166
Pennell v. Payne [1995] 2 All ER 592 32
Pochin v. Smith [18887] 52 JP 4 185
Pointe Gourde Quarrying and Transport Company Limited v. Sub-
  Intendent of Crown Lands [1974] AC 565 128
Pole v. Peake [1998] EGCS 125 166
Polkey v. AE Dayton Services Ltd [1988] ICR 142 246
Polsue and Alfieri v. Rushmer [1970] AC 121 164
Prassad v. Wolverhampton Borough Council [1983] JPL 449 132

Table of Cases 269

Pride of Derby Angling Association *v.* British Celanese [1953]
1 All ER 179 193
Punchknowle Farms Ltd *v.* Kane [1985] 3 All ER 790 226

R *v.* Oxfordshire CC ex parte =Sunningwell Parish Council [1999]
3 WLR 160 84
R *v.* Pratt (1885) 4 E&B 860 187
R *v.* Teignbridge District Council Ex parte Street, *The Times* 9 October
1989 172
R *v.* Whittaker (1848) 17 LJMC 127 190
Ratcliffe *v.* McConnell [1999] 1 WLR 670 161
Rawson *v.* Peters (1973) EGD 259 193
Read *v.* Edwards (1864) 17 CBNS 245 200
Read *v.* Lyons & Co. Ltd [1945] KB 216 163
Rees *v.* Morgan (1976) 240 EG 787 171
Riordan Communications Ltd *v.* South Bucks DC [1999] EGCS 146
.... 99
Robinson *v.* Kilvert (1899) 41 ChD 88 164
Robinson *v.* Vaughan (1838) 8 C&P 252 200
Rodd *v.* Ritchings [1995] RA 299 261
Rugby Joint Water Board *v.* Walters [1966] 3 All ER 497 175
Rutherford *v.* Maurer [1961] 2 All ER 755 31
Rylands *v.* Fletcher (1868) LR 3 HL 330 .... 168, 169

St Helens Smelting Co. Ltd *v.* Tipping (1865) 11 HL Cas. 642
.... 164
St John's College Oxford *v.* Thames Water Authority [1990]
1 EGLR 229 155
Scott-Whitehead *v.* National Coal Board (1987) 52 P&CR 263
.... 175
Sedleigh-Denfield *v.* O'Callaghan [1940] AC 880 163
Seligman *v.* Docker [1948] 2 All ER 887 165
Skerritt's of Nottingham *v.* Secretary of State [2000] 20 PELB 90
Skerritt's of Nottingham *v.* Secretary of State [2000] EGCS 43 94
Smeaton *v.* Ilford Corporation [1954] Ch.450 168
Smith *v.* Andrews [1891] 2 Ch. 678 195
Solloway *v.* Hampshire County Council (1981) 79 LGR 449 166
Spicer *v.* Bernard (1859) 23 JP 311 191
Stokes *v.* Cambridge City Council [1961] 13 P&CR77 126

Taylor & another *v.* North West Water Ltd [1995] 1 EGLR 266
.... 155
The Wagon Mound (No.2) [1967] AC 667 167
Thomas *v.* Witney Aquatic Co. Ltd (1972) RRC 348 260
Tillett *v.* Ward (1882) 10 QBD 17 171
Titchener *v.* British Railways Board [1983] 3 All ER 770 161

Wagstaff v. Secretary of State DETR [1999] 21 EG 137 119
Wallace v. Newton [1982] 2 All ER 106 170
Walsingham's case 1
Watkins v. Hereford Assessment Committee (1935) 154 LT 262
   .... 257, 260
Welsh National Water Development Authority v. Burgess (1974)
   RVR 395 .... 182, 193
West Riding of Yorkshire River Board v. Tadcaster UDC (1907)
   97 LT 436 194
Wharton v. Taylor (1965) 109 SJ 475 (DC) 198
Wheeldon v. Burrows (1879) 12 Ch 31 4
Wildtree Ltd v. Harrow LBC [2000] EGCS 80 133
Womersley (VO) v. Jisco [1990] RA 211 260
Wykes v. Davis [1975] 1 All ER 399 38

Table of Statutes

Acquisition of Land Act 1981 .... 119, 147, 154
   s.151 123
   s.154 150
   s.159 .... 150, 151
   s.160 .... 150, 151
   s.163 151
   s.168 150
   s.175 .... 150, 151
   Sch.1 Para.3(1)(a) 123
   Sch.2 147
   Sch.13 123
   Sch.18 151
   Sch.19, Para.7 151
   Sch.21 151
      para.2 151
      para.4 151
      para.5 151
Agricultural Holdings Act 1986
.... 12, 18, 19, 31, 34, 48, 60, 65, 155, 230, 231
   s.1(2) 31
   s.1(3) 32
   s.2 .... 36, 46
   s.3 36
   s.4 36
   s.6 .... 32, 46, 229
   s.7 .... 33, 46
   s.7(2) 41
   s.8 .... 33, 46
   s.9 46
   s.10 33
   s.11 33
   s.12 .... 34, 35, 46
   s.13 .... 35, 46
   s.14 35
   s.15 35
   s.20 .... 41, 46, 165, 186
   s.25 36
   s.25(3) 36
   s.25(4) 36
   s.26 .... 36, 37
   s.27 .... 37, 44

s.28 39
s.35 43
s.36(2) 42
s.36(4) 43
s.37 .... 42, 45
s.37(1)(b) 44
s.38(4) 42
s.38(5) 42
s.39 43
s.39(6) 43
s.42 43
s.44 44
s.45 44
s.46 44
s.47 44
s.48(6) 44
s.49 45
s.49(1)(b) 45
s.50 45
s.51(2) 45
s.51(4) 45
s.51(5) 45
s.52 45
s.53(8) 45
s.57(4) 45
s.60 41
s.61 41
s.66(1) 40
s.68 40
s.70 41
s.71 41
s.83 .... 39, 46
s.84 46
s.96 31
Sch.1 32
Sch.2 .... 34, 45
   Part IV 42
Sch.3 .... 38, 46
   para.10 38
Sch.6
   para.2 43
   para.3 43
   para.4 43
   para.6(12)(d) 43
Sch.7 39
   Part I 39, 40

Table of Statutes

Part II	40
Sch.8	39
Part II	40
Sch.11	46

Agricultural Tenancies Act 1925
- s.6 18
- s.7 18

Agricultural Tenancies Act 1995
  .... 4, 12, 13, 14, 16, 18, 23, 60, 230, 231
- Part II 22
- Part III 28
- s.1 12
- s.5 17
- s.6 17
- s.7 18
- s.7(3) 18
- s.8 23
- s.9(a) 19
- s.9(b)(1) 19
- s.9(b)(ii) 19
- s.10 .... 20, 21, 22, 27
- s.11 21
- s.12 .... 22, 28
- s.13 .... 21, 22
- s.15 24
- s.16(1) 23
- s.17 .... 23, 24, 28
- s.18(1) 25
- s.19 .... 25, 28
- s.19(1) 25
- s.19(10) 24
- s.20 26
- s.21 26
- s.22(1) 26
- s.22(2) 26, 28
- s.22(3) 28
- s.23 26
- s.24 27
- s.25 27
- s.28 29
- s.28(4) 29
- s.29 29
- s.36 .... 29, 30
- s.36(7) 30
- s.38(2) 12

Agricultural Tenancies Act 1999 .... 233, 234

s.15 233
Agricultural Wages Act 1948 48
Agriculture Act 1947 .... 165, 189
   s.98 165
Agriculture Act 1986 .... 232, 234
   s.15 231
   Sch.1 233
     para.1 232
     para.1–4 232
   Sch.10, Para.10 233
Allotments Act 1922 257
Ancient Monuments and Archaeological Areas Act 1079, s.1 256
Animals Act 1971 .... 80, 170
   s.2(1) 169
   s.2(2) 169, 170
   s.3 171
   s.4 170
   s.5(5) 171
   s.6(2) 169
   s.7 170
   s.8 80
   s.8(1) 171
   s.11 171
Arbitration Act 1996 46

Caravan Sites and Control of Development Act 1960 91
Clean Air Act 1956
   s.1 210
   s.11–s.15 210
Commons Registration Act 1965
   s.7 172
   s.22(1) 84
Companies Act, s.319 238
Compulsory Purchase Act 1965 .... 118, 125, 147
   s.7 130
   s.10 133
   s.10(2) 119
Control of Pollution Act 1974
   Part I 219
   Part III 163
Countryside Act 1968, s.30 70
Countryside and Rights of Way Act 2000
   .... 69, 71, 72, 77, 80, 83, 84
   s.13(3) 83
   s.20 83
   s.39 72

s.47	70, 75
Criminal Justice and Public Order Act 1994	172
s.61	172
s.63	172
s.68	172
s.70	173
Criminal Justice and Public Order Act 1997, s.68	200
Criminal Law Act 1970	206
Dangerous Dogs Act 1989	79, 170
Dangerous Dogs Act 1991	79, 169
Dangerous Wild Animal Act 1976	169
Deer Act 1991	192, 202, 203, 204, 205, 206
s.12	202, 203
s.13	203
Defective Premises Act 1972	162
s.4(1)	162
s.4(2)	162
Disability Discrimination Act 1995	238
Dogs Act 1871	79, 170
Electricity Act 1989	138, 139, 141, 143
Sch.3	141
Sch.4	141
para.3	143
para.8	142
para.9	144
para.10	143
Employers' Liability (Compulsory Insurance) Act 1969	158
Employment Relations Act 1999	250
Employment Rights Act 1996	241
s.8	240, 243
s.13	243
s.86	240, 243
s.94	244
Environment Act 1995	178, 200
s.6(6)	200
s.13	200
s.40	182
s.41(1)(a)	178
s.41(8)	178
s.53	182
s.57	211
Environmental Protection Act 1990	
Part I	209, 210
Part II	218, 219

Part IIA	211
Part III	162
s.33	216, 217
s.34	216, 222
s.35	221
s.36	222
s.37–s.39	222
s.43	222
s.59	217
s.62	220
s.74	221
s.75	219
s.75(2)	218
s.75(3)	219
s.75(7)	220
s.75(7)(c)	218
s.78A–s.78YC	211
s.78A(2)	212
s.78A(4)	213
s.78A(9)	213
s.78B	213
s.78E(1)	214
s.78F	215
s.78H	214
s.78L	215
s.79	162
s.152	166
Sch.22	216
Equal Pay Act 1970	239
Finance Act 1931, s.28	67
Firearms Act 1968	202, 203, 204, 205, 206
s.1	192, 205
s.2(1)	205
s.10	206
s.17	205
s.19	78, 191, 202, 203
s.19A	202
s.20	202, 203
s.20(2)	191, 206
s.22(3)	192
s.27	192
s.46	202, 204
s.47	203
s.47(2)	206
s.47(4)	203

s.48	202, 203
s.48(2)	204
s.52	203
Sch.6, Part.1	205
Firearms Act 1997	191
Firearms Acts 1968–1988	190
Firearms (Amendment) Act 1988	191
s.1(2)	192
s.2	192
s.3	192
Food and Environment Protection Act 1985, Part III	210
Game Act 1831	190, 202, 203, 204, 205, 206
s.3	189, 205
s.12	191
s.13	203
s.18	189
s.23	189
s.30	190, 205
s.31	202, 203
s.31A	202
s.32	205
s.35	200
s.36	203
Game Laws (Amendment) Act 1960	190, 202, 203, 204, 205, 206
s.2	202
s.3	203
s.4	203, 204
Game Licence Act 1860	
s.4	189, 205
s.14	189
Gas Act 1986	139, 140, 146, 147, 156
s.9	147
Sch.2	139
Sch.3	147
Gas Act 1996	149
General Rate Act 1967	254
Ground Game Act 1880	185, 186, 189
s.1	186
s.6	185
Guard Dogs Act 1975	79, 170
Health and Safety at Work Act 1974	74, 161, 237, 253
s.3	79
s.15	161
s.47	161

Highways Act 1980 .... 71, 74, 75, 77
   s.26 83
   s.30 217
   s.31 72
   s.31(6) 85
   s.134 75
   s.137 72
   s.148 216
   s.154 166
   s.155 80
   s.161 78
   s.164 74
Housing Act 1980 58
   s.89 59
Housing Act 1988 .... 47, 52, 53, 54, 57, 58, 62
   s.5 55
   s.13 .... 55, 58
   s.20 58
   s.21 59
   s.22 .... 53, 58
   s.25–s.32 18
   s.34 53
   Sch.2 54
   Sch.3 53
Housing Act 1996 .... 58, 59
   s.96 58
   s.97 58
   Sch.7 58
Human Rights Act 1998 .... 96, 110, 116, 118, 151, 154
   Art.6 122

Land Charges Act 1972 .... 5, 7
Land Clauses Consolidation Act 1845 118
   s.68 119
Land Compensation Act 1961 .... 118, 125
   s.4 129
   s.5 147
      Rule.6 131
   s.7 130
   s.17 127
Land Compensation Act 1965 .... 150, 154
Land Compensation Act 1973 .... 119, 125
   Part.1 133
Land Drainage Act 1991 .... 151, 174, 181
   Part IV, Chapter II 183
   s.14(5) 182

s.15	183
s.15(1)(b)	183
s.22	152, 184
s.25	152, 184
s.28	152
s.28–s.29	183
s.29(3)–(7)	183
s.30	183
s.33	184
s.62	152, 182
s.69(1)	182

Land Registration Act 1925 10
Land Registration Act 1988 6
   s.1(1) 7
Landlord and Tenant Act 1927 65
   s.1(1) 66
Landlord and Tenant Act 1954 .... 14, 15, 16, 66, 141
   Part II 13, 60, 62
   s.23(1) 13, 62
   s.23(2) 13
   s.24–s.28 67
   s.38(4) 67
Landlord and Tenant Act 1985, s.11 50
Landlord and Tenant Act 1987
   s.47 38
   s.48 .... 30, 38
Law of Property Act 1925
   s.1 3
   s.62 4
   s.99 4
   s.140 18
   s.149(6) 18
   s.193 81
Limitation Act 1980 7
Local Government Finance Act 1988 254
   s.45 256
   s.46 256
   s.46A 256
   s.47 256
   s.48 256
   s.64(4) 257
   s.64(11) 257, 259
   s.66 254
   Sch.5 257
      para.2(1) 257
      para.2(2) 257

para.3	258
para.4	258
para.5	259
para.6	259
para.7	258
para.8(3)	258
para.9	259
Sch.6, para.2	254
Local Government Finance Act 1992	261
s.12	264
s.66(2A)	261
Local Government and Rating Act 1997	255, 256
s.2	259
Mines and Quarries Act 1954	78
National Minimum Wage Act 1998	243
National Parks and Access to the Countryside Act 1949	81
New Roads and Street Works Act 1991	139
s.48(1)	138
Night Poaching Act 1828	190, 202, 203, 204, 205, 206
s.1	202, 203, 205
s.2	202, 205
s.9	203, 205, 206
Night Poaching Act 1844	190, 202, 205, 206
s.1	205
Occupiers' Liability Act 1957	77, 82, 159, 162
s.2(2)	160
s.2(3)(a)	160
s.2(3)(b)	160
s.2(4)(a)	160
s.2(4)(b)	160
s.2(6)	159
Occupiers' Liability Act 1984	77, 82, 159, 161
s.1(3)	161
s.1(4)	161
s.1(5)	161
s.1(6)	159
s.1(7)	159
s.1(8)	161
s.1(9)	161
s.2	160
Pests Act 1954	165
Pipelines Act 1962	139, 147, 156

s.9	157
s.11	157
Sch.2	157
Planning and Compensation Act 1991	126
Sch.12	99
Planning (Listed Buildings and Conservation Areas) Act 1990	
s.1(1)	104
s.6	104
s.9	105
Poaching Prevention Act 1862	190, 203, 204, 205, 206
s.2	203, 205
Police and Criminal Evidence Act 1984	199, 202, 203, 204
Part III	191
s.25	202, 203
Sch.6	203
Prescription Act 1832	194, 195
Protection of Animals Act 1911	188
s.1	188
Protection of Animals (Amendment) Act 2000	188
Protection of Badgers Act 1992	188
s.10	188
Protection From Eviction Act 1977	57
s.5	18
Public Health Act 1961, s.34	218
Race Relations Act 1976	238
Railway Clauses Consolidation Act 1845	132
Rating (Former Agricultural Premises and Rural Shops) Act 2001	255
Refuse Disposal (Amenity) Act 1978	
s.2	216
s.3–s.5	217
s.6	218
Rehabilitation of Offenders Act 1974	239
Rent Act 1977	47, 49, 51, 56, 57, 58
Sch.15	56
Rent (Agriculture) Act 1976	47, 54, 56, 57
s.1	48
s.2(2)	48
s.2(4)	49
s.4	52
s.4(1)	49
s.6	51
s.7	51
s.10	50
s.11	50

s.12 .... 50, 53
s.27(2) 56
Sch.2, para.1 49
Sch.3 .... 47, 49
Sch.4 51
Sch.5 50
Requisitioned Land and War Works Act 1948 156
Rights of Way Act 1990 .... 74, 75, 76

Salmon Act 1986 199
Salmon and Freshwater Fisheries Act 1975 200
s.1 198
s.1(1) 198
s.1(4) 198
s.2 .... 196, 197, 198
s.3 198
s.19 196
s.19(3)–(5) 197
s.19(6) 197
s.19(7) 197
s.19(8) 197
s.20 197
s.22 197
s.23 197
s.25(2) 198
s.25(4) 197
s.25(5) 197
s.25(6) 197
s.25(7) 198
s.26 198
s.27 197
s.30 181
s.35 198
s.41 196
Sch.1 197
  para.3 196
Sch.2 197
  para.15 198
  para.16 197
  Part IV 197
Sch.4 181
Settled Land Act 1925 7
Sex Discrimination Act 1975 238
Sex Discrimination Act 1986 238
Statutory Water Companies Act 1991 174

Table of Statutes 283

Telecommunications Act 1984 .... 138, 143
   s.34 144
   Sch.2 144
      para.1 139
The Deer Act 1991 188
Theft Act 1968 199
   s.4(4) 190
   s.32(1) 199
   Sch.1, para.2 199
Torts (Interference with Goods) Act 1977
   s.12 170
   s.13 170
   Sch.1 170
Town and Country Planning Act 1971, s.52 99
Town and Country Planning Act 1990 .... 111, 119
   s.54A 86, 116
   s.55(1) 87
   s.55(2) 87
   s.55(2)(e) 103
   s.70 86
   s.70(1) 97
   s.73 97
   s.78 109
   s.106 .... 97, 98, 99
   s.171B 113
   s.172 112
   s.183 113
   s.191 113
   s.192 114
   s.197A 111
   s.215–s.219 218
   Sch.13 para.21 123
   Sch.13 para.22 123
Town Police Clause Act 1847 .... 78, 79
Transport and Works Act 1992 .... 95, 120
Trusts of Land and Appointment of Trustees Act 1996 3

Unfair Contract Terms Act 1977 .... 160, 241
   s.11(3) 160

Water Act 1945 155
Water Act 1973 175
Water Act 1989 .... 174, 175, 180, 200
Water Industry Act 1991 .... 139, 150, 152, 174, 175
   s.12 157
   s.41 .... 152, 154

s.98 154
s.117 153
s.155 175
s.158 154
s.159 .... 154, 155
s.165 155
s.179 155
s.180 156
Sch.11 154
Sch.12 156
Water Resources Act 1963 180
Water Resources Act 1991 .... 139, 149, 154, 174, =176, 181
   Part II 176
   s.19 180
   s.24 177
   s.25 180
   s.25(3) 180
   s.25(8) 180
   s.27 149
   s.27(1) 177
   s.27(2) 177
   s.27(3) 177
   s.27(4) 177
   s.27(5) 177
   s.27(7) 177
   s.29(1) 177
   s.29(2) 177
   s.32(2) 177
   s.32(3) 177
   s.32(4) 177
   s.33 177
   s.35 177
   s.37 178
   s.38 178
   s.39 179
   s.40 178
   s.43–s.45 179
   s.46 178
   s.48 179
   s.49 179
   s.51 179
   s.52–s.53 179
   s.57 180
   s.60 179
   s.65 177
   s.73–s.81 180

Table of Statutes

s.85	207, 208, 211
s.86	208
s.88	208
s.92	208
s.93	209
s.94	209
s.95	209
s.97	211
s.107	184
s.113(1)	181
s.154	175
s.193	181
s.194	181
s.221	176
Sch.7	177
Sch.10	208
Sch.21, para.5(1)	182
Weeds Act 1959	167
Wild Creatures and Forest Laws Act 1971	202
Wild Mammals (Protection) Act 1996	188
Wildlife and Countryside Act 1981	69, 187
s.1–s.8	187
s.1(5)	188
s.2(3)	188
s.2(4)	188
s.2(5)	188
s.2(6)	188
s.2(7)	188
s.3	188
s.4(1)	187
s.4(2)	187
s.4(3)	187
s.5	193
s.9–s.12	188
s.11	193
s.14	181
s.27(1)	188
s.39	79
s.66(1)	70
Sch.1	187, 188
Part I	187
Part II	188
Sch.2	187, 188
Part I	188
Part II	188
Sch.5–7	188

Sch.7 186
　　　Sch.9 181

Zoo Licensing Act 1981 169

Table of Statutory Instruments

Agricultural Wages Order 48, 240, 242, 250
Agriculture (Calculation of Value for Compensation) (Amendment)
  Regulations 1981 (SI 1981/22) 40
Agriculture (Calculation of Value for Compensation) (Amendment)
  Regulations 1983 (SI 1983/1475) 40
Agriculture (Calculation of Value for Compensation) Regulations 1978
  (SI 1978/809) 40
Agriculture Holdings (Arbitration on Notices) Order 1987 (SI 1987/710)
  .... 36, 37
Agriculture (Maintenance, Repair and Insurance of Fixed Equipment)
  Regulations 1973 (SI 1973/1473) 33
Animal By-Products Order 1999 (SI 1999/646) 221

Clean Air (Emission of Dark Smoke Exemption) Regulations 1969
  210
Compulsory Purchase by Ministers (Inquiries Procedure) Rules 1994
  (SI 1994/3264) 122
Compulsory Purchase by Non-Ministerial Acquiring Authorities
  (Inquiries Procedure) Rules 1990 (SI 1990/512) 122
Conservation (Natural Habitats, etc.) Regulations 1994 (SI 1994/2716)
  107
Control of Pollution (Silage, Slurry and Fuel Oil) Regulations 1991
  (SI 1991/324 as amended) 208
Control of Substances Hazardous to Health Regulations 1999 (SI1999/)
  253
Council Tax (Chargeable Dwellings) Order (SI 1992/549) 261
Council Tax (Prescribed Classes of Dwellings)(Wales) Regulations 1998
  (SI 1998/105) 264
Crop Residues (Burning) Regulations 1993 (1993/1366) 166

Dairy Produce Quotas Regulations 1997 (SI 1997 No 733)
  .... 224, 226, 227, 228, 230, 233

Electricity Generating Stations and Overhead Lines (Inquiries Procedures)
  Rules 1990 (SI 1990/528) 141
Environment Protection (Prescribed Processes and Substances)
  Regulations 1992 (SI 1991/472 as amended by SI 1992/614)
  209
Environmental Protection (restriction on Use of Lead Shot), (England)
  Regulations 1999 (SI 1999/2170) 193

Gas (Street Works)(Compensation for Small Businesses) Regulations 1996 149

Hedgerow Regulations 1997 (SI 1997/1160) 223

Land Registration Rules 1925 10
Landlord and Tenant Act 1945, Part II (Notices) Regulations 1983 (SI1983/133) 63
Landlord and Tenant Act 1945 (Appropriate Multiplier) Order 1990 (SI 1990/363) 65

Noise at Work Regulations 1989 (SI 1989/1790) 253
Non-Domestic Rating (Former Agricultural Premises)(England) Order 2001 (SI 2001/2585) 255
Non-Domestic Rating (Stud Farms)(England) Order 2001 (SI 2001/2586) 259
Non-Domestic Rating (Unoccupied Property) Regulations 1989 (SI 1989/2261, as amended by SI 1995/549) 256

Part-Time Workers (Prevention of Less Favourable Treatment) Regulations 2000 239
Protection of Water against Agricultural Nitrate Pollution Regulations 1996 (SI 1996/888) 209

Rent Acts (Maximum Fair Rent) Order 1999 50
Reporting of Injuries, Diseases and Dangerous Occurrences Regulations 1995 (SI 1995/) 253

Sludge (Used in Agriculture) Regulations 1989 (SI 1989/1263, as amended) 211, 221
Statutory Nuisance (Appeals) Regulations 1995 (SI 1995/2644) 162

Town and Country Planning (Control of Advertisement) Regulations 1992 (SI1992/666) 87
Town and Country Planning (Demolition – Description of Buildings) Direction 1995 87
Town and Country Planning (Environmental Impact Assessment) (England and Wales) Regulations 1999 (SI 1999/293) 107
Town and Country Planning (Fees for Applications and Deemed Applications)(Amendment) Regulations 1997 (SI 1997/37) 97
Town and Country Planning (General Development Procedure) Order 1995 (SI 1995/419) 96
Town and Country Planning (General Permitted Development) Order 1995 (SI 1995/418) 88, 89, 90, 91, 92, 93, 94, 95, 96, 107

Town and Country Planning (Trees) Regulations 1999 (SI 1999/1892) 106
Town and Country Planning (Use Classes) Order 1987 (SI 1987/764) 90
Transfer or Undertakings (Protection of Employment) Regulations 1981 .... 251, 252

Waste Management Licensing Regulations 1994 (SI 1994/1056) 219

Index

abandoned vehicles 217
abandonment, planning control and 103
abatement notices 162–3
absolute title 9
abstraction licences 176–80
access rights *see* public access
accommodation works, compensation for 132
advance payments, compulsory purchase compensation 129
adverse possession 7
Advisory, Conciliation and Arbitration Service (ACAS) 235, 236, 244
agents, service of notices on 29–30
aggravated trespass 172–3
agreement
  agricultural holdings tenancy agreement 32
  for a lease 7
  purchase of land by 'agreement' 124
agricultural buildings and operations
  planning control 92–4
  rating 258–9
Agricultural Dwellinghouse Advisory Committees (ADHACs) 52, 56
agricultural holdings 31–46
  business tenancies 60
  compensation 39–41
  definitions 31–2
  disputes procedure 45–6
  fixed equipment 32–3
  milk quotas and 230
  rent 34–5, 44–5
  security of tenure 36–9
  sporting rights 185–6
  succession tenancies 42–5
  tenancy agreement 32
Agricultural Lands Tribunal 35, 40, 46, 183
agricultural occupancy condition in grant of planning permission 100–102

agricultural tenancies *see* farm business tenancies
air pollution 210, 211
ancient monuments 107, 108
animals
  buildings for 258–9
  dangers 79–80, 169
  nuisance caused by 165–6
  protection of wildlife 188
  rights of way and 79–80
  straying 169–71
appeals
  abstraction licences 179
  to Employment Appeal Tribunal (EAT) 236
  planning control 108–111
    judicial review 109–110
    procedure 108–109
    third party rights 110–111
  remediation notices 215
appropriate alternative development, certificate of 127
arbitration
  compensation for improvements and 25
  dispute resolution and 46
  failure to agree and 28
  rent 21, 22–3, 34, 35, 44–5
  tenancy agreements 32
areas of outstanding natural beauty (AONB) 95, 107, 108
armed trespass 191
arrest, powers of 191, 202
assured agricultural occupancy 53–4
assured shorthold tenancies 58

bad husbandry certificate 35
badgers 188
bankruptcy 237
beekeeping 259
betterment, compulsory purchase and 127–8, 130–131
birds, protection of 187–8
blight notice procedure 123–4

Index

bona vacantia 2
boundaries 8
breach of condition notices 111
breach of covenants 17, 19
breach of employment contract 235, 242
break clauses 18
bridleways 70
brother, succession of tenancies to 42
buildings
   building preservation notice 104
   demolition 87, 105
   listed 90, 95, 103, 104–106
   rating and 258–9
   see also planning control
bulls 79, 169
business tenancies 13, 15, 60–68
   agricultural holdings 60
   compensation 65–7
   farm business tenancies 60–61
   Landlord and Tenant Act 1954 62–7
   legal nature of lease or tenancy 61–2
   registration 68
   rent 64–5
   stamp duty 67–8
   termination 63
byways 70

cable companies 146
caravans, planning control 91–2
caution, minor interests and 7–8
certificates
   of appropriate alternative development 127
   bad husbandry certificate 35
   charge certificate 8
   for firearms 191–2
   land certificate 8
   lawful development certificates 113–14
change of landlord 30
change in ownership of business 251–3
changes of use 90–91
charge certificate 8
charges register 9
charities, rating and 256
children, succession of tenancies to 42–3
close seasons, fishing 196–7

clubs, rating and 256–7
cohabitees, succession of tenancies to 52
commercial equitable interests 5
commercial waste 220
common land 69, 80–84
companies, service of notices on 30
company directors 238
compensation
   compulsory purchase 124–35
      accommodation works 132
      advance payments 129
      betterment 127–8, 130–131
      disturbance compensation 131–2, 133
      entitlement 124–5
      equivalent reinstatement 128
      interest 129–30
      land taken 125–7
      miscellaneous matters 134–5
      planning assumptions 127
      settling claim 129
      severance and injurious affection 130
      use of public works 133–4
      valuation date 128–9
   dilapidation 41
   disturbance 40–41, 65, 131–2, 133
   game damage 41, 185–6
   gas pipeline works in streets 149
   improvements
      agricultural holdings 39–40
      business tenancies 65–7
      farm business tenancies 23–7, 28
   milk quotas and 232–4
   pipelines 147
   unfair dismissal 246–7
   water resources and drainage 151
   water supply and sewerage 155–6
   wayleaves 142–3, 145–6
compulsory purchase 118–36
   compensation 124–35
      accommodation works 132
      advance payments 129
      betterment 127–8, 130–131
      disturbance compensation 131–2, 133
      entitlement 124–5
      equivalent reinstatement 128
      interest 129–30

for land taken 125–7
miscellaneous matters 134–5
planning assumptions 127
settling claim 129
severance and injurious affection 130
for use of public works 133–4
valuation date 128–9
Critchel Down rules 135–6
electricity suppliers 141, 142
gas pipelines 147–8
legislative background 118–19
pipelines 157
powers to acquire land 120–124
blight notice procedure 123–4
exercise of CPO 122–3
inquiry 121–2
missing owners 124
procedures 120–121
purchases by 'agreement' 124
water resources and drainage 150–151
water supply and sewerage 154–6
compulsory rights order (CRO) 157
compulsory works order (CWO) 150, 151
consent
landlord's consent to improvements 28, 39
landlord's consent for planning permission 24–5
conservation areas, planning control 95, 106–107
consultation with trade unions 250
contaminated land 211–16
buying and selling 216
civil liability 216
definition 212–13
local authority inspection 213–14
register 215
remediation notices 214–15
work notice procedures 216
contract of employment 235, 239
breach 235, 242
changes to 242
implied terms 241–2
written statement of employment 239–41
corporeal hereditaments 2
council tax 261–4

courts, dispute resolution and 29, 46
covenants
breach of 17, 19
restrictive 5–6, 7
criminal offences
waste disposal 222
water pollution 207–208
criminal records 239
Critchel Down rules 135–6
crop spraying, rights of way and 74
Crown
land and 1–2
foreshore 187, 194
registration 6–7

damage
caused by game/wild animals 41, 165–6, 185–6
straying animals 169–70
damages 167
dangers
animals 79–80, 169
rights of way and 77, 78–9
animals 79–80
death
redundancy and death of employer 249
succession of tenancies on 42–5
deer 188, 192–3
demolition
planning control 87
listed buildings 105
development plan 114–17
dilapidation, compensation for 41
directors of company 238
discrimination 238–9
dismissal 244–7
dispute resolution
agricultural holdings 45–6
farm business tenancies 27–9
disturbance compensation 40–41, 65, 131–2, 133
diversion order 71
dogs
dangerous 169, 170
exclusion from open land 81–2
rights of way and 79–80
worrying livestock 171
drainage rights 4
drought orders 180

Index 293

dwellinghouses
  council tax 261–4
  planning control 88–90
  see also residential protection of farmworkers

easements 3–4, 7, 137, 142
electricity suppliers 139, 140–144
employment issues
  breach of contract 235, 242
  change in ownership of business 251–3
  choice of employed or self-employed status 236–8
  disciplinary procedures 244
  dismissal 244–7, 252
  duty owed to employees 161, 241–2
  Employment Tribunals 235–6
  engaging a worker 238–42
  health and safety at work 237, 253
  notice for termination of employment 240, 243–4
  redundancy 247–51
  rights under employment legislation 242–4
enforcement
  enforcement notices 105, 112–13
  planning control 105, 111–14
entry, notice of 122
Environment Agency (EA) 149, 174–5, 178–9, 181–3, 200
environmental impact assessment (EIA) 107
environmental protection 207–23
  air pollution 210, 211
  codes of practice 210–11
  contaminated land 211–16
    buying and selling 216
    civil liability 216
    definition 212–13
    local authority inspection 213–14
    register 215
    remediation notices 214–15
    work notice procedures 216
  fly tipping 216–18
  hedgerows 223
  integrated pollution control 209
  pesticides 210
  soil protection 211
  waste disposal 218–23

    definitions 218–20
    duty of care 222–3
    exemptions 220–221
    offences 222
    waste management licence 221–2
  water pollution 207–209
    offences 207–208
    preventative measures 208, 211
    protection areas 209
environmental study (ES) 107
equitable interests in land 4–6
escape, strict liability for 168–9
estates in land 1, 2–3
examination in public (EIP) 115
extinguishment order 71

family equitable interests under trust 4–5
farm business tenancies 12–30, 60–61
  agriculture condition 14
  arbitration 21, 22–3
  business conditions 12–14
  compensation for improvements 23–7, 28
  dispute resolution 27–9
  effect of breach of covenant 17, 19
  fixtures 23
  meaning of 12–17
  milk quotas and 230, 233–4
  notices 14–16, 29–30
  rent 19–23
  termination of 17–19
farm loss payment 135
farmworkers, residential protection see residential protection of farm workers
fee simple 1, 2
fees, planning permission 97
fences 73–4
feudal system 1
fire
  air pollution caused by 210, 211
  nuisance caused by straw and stubble burning 166
  rights of way and 78
firearms 187, 191–3
  armed trespass 191
  rights of way and 78
fish farms and fisheries, rating of 259–60

fishing 193–9
  close seasons 196–7
  illegal methods 198–9
  leases for 195
  licences for 197–8
  ownership of fishing rights 193–5
  poaching of fish 199
  protective fishing rights 193
  public 195
  stocking of fishing lakes 181
  unclean and immature fish 197
fixed rent 19
fixed term tenancies, termination of 17, 61
fixtures
  agricultural holdings 32–3
  farm business tenancies 23
flood defence 181–4
fly tipping 216–18
footpaths 70
forestry buildings and operations, planning control 94
forfeiture 19
freehold 2, 9
fuel stores 208

game
  damage caused by 41, 165–6, 185–6
  game laws 189
  hunting 200–201
  occupier's right to ground game 186
  poaching of 189–91
gas 139, 146–9
gates 73–4
general vesting declaration 122
government guidance, planning control 115, 117
grazing agreement, milk quotas and 228–9
grievance procedures 240–241
ground game, occupier's right to 186
groundwater 176
guard dogs 79

health and safety at work 237, 253
hedgerows 223
high farming 41
highways
  animals straying onto or off 170–171
  fly tipping 216–17

utilities works in highway land and private tracks 138–40, 149
see also rights of way
home loss payment 134
horse grazing 102–103
household waste 219–20
housing see dwellinghouses; residential protection of farmworkers
hunting 200–201
husbandry, agricultural holdings and 35

illegal methods of fishing 198–9
impounding water 180–181
improvements
  compensation for
    agricultural holdings 39–40
    business tenancies 65–7
    farm business tenancies 23–7, 28
  landlords, rent increase and 35
  landlord's consent to 28, 39
incorporeal hereditaments 2
indemnity, registration of title and 10
industrial waste 220
inhibitions, minor interests and 8
injunctions 167
injurious affection, compulsory purchase and 130
inquiries
  abstraction licences 179
  compulsory purchase schemes 121–2
insolvency, redundancy payments and 251
insurance 158
  obligations for agricultural holdings 32–3
integrated pollution control 209
interest, compulsory purchase compensation 129–30
interim rent 64–5
irrigation 180

joint tenants 1
judicial review, planning control appeals 109–110

land certificate 8
land law, general principles 1–2
landlords
  change of landlord 30

Index 295

consent
    to improvements 28, 39
    for planning permission 24–5
  improvements and 35
  occupier's liability 162
Lands Tribunal, application for discharge of restrictive covenant 6
lawful development certificates 113–14
lead shot ban 193
leases and leasehold 2–3
  lease for lives 18
  legal nature of lease 61–2
  overriding interests 7
  types of title 9
  see also agricultural holdings; business tenancies; farm business tenancies
legal estates in land 2–3
legal interests in land 3–4
licences 61
  abstraction licences 176–80
  fishing 197–8
  shotgun licence 191–2
  waste management licence 221–2
listed buildings
  planning control 90, 95, 103, 104–106
    listed building consent 105
lives, lease for 18
livestock *see* animals
local authorities
  contaminated land and 212, 213–14
  council tax 261–4
  fly tipping and 217, 218
  local plan 114, 115–16, 127
  rating 254–61
  see also planning control

maintenance and repair obligations, agricultural holdings 32–3
merger 19
milk quotas 224–34
  leasing of quota 230
  legal status 229–30
  special quota 225
  tenancies and 230–234
  transfer of quota 225–30
    with land 225–8
    particular questions 228–9

    without land 228
minerals
  compulsory purchase and 135
  gas pipelines and 147–9
  mineral planning guidance notes (MPGs) 117
  mineral plans 116
mines 78
minor interests 7–8
minor works, planning control 90
mixed uses, rating and 257–8
modification orders 70
mortgages 3, 4
  charge certificate 8

National Grid Company (NGC) 140–141
National Parks 95, 107, 108
negative covenant 5
negligence 159
nitrate sensitive areas (NSAs) 209
noise 163
non-visitors, duty owed to 161
Norfolk Broads 95, 108
notice
  farm business tenancies and 14–16, 29–30
  minor interests and 7
  notice of redundancies 250
  to pay rent 38
  to remedy breach of tenancy 38–9
  service of 29–30
  termination of employment 240, 243–4
  termination of tenancies
    agricultural holdings 36–8
    farm business tenancies 17–19
nuisance 162–8
  contaminated land as 216
  private 163–8
    definition 163
    main rules 163–5
    remedies 167–8
    types of 165–7
  public 162
  statutory 162–3

obstruction of highway 72–3
occupier's liability 158–73
  duty owed to employees 161

duty owed to non-visitors 161
duty owed to visitors 159–60
  agreements changing the duty 160–611
  insurance 158
  landlords and 162
  legislation on 159
  negligence 159
  nuisance 162–8
    private 163–8
    public 162
    statutory 162–3
  straying animals 169–71
  strict liability for escape 168–9
  trespass 161, 171–3
occupier's right to ground game 186
open land 80–84
opposition to new tenancy, grounds for 63–4
option to purchase the reversion 7
oral tenancies 32
overhanging vegetation, rights of way and 74
overreaching 5
overriding interests 7
ownership of business, change in 251–3

pay 243
penalties
  contaminated land 215
  poaching 205–206
pensions 240
periodic tenancies, termination of 17–18, 61
pesticides 210
phased rent 19
pipelines
  gas 139, 146–9
  other pipelines 139, 156–7
  pipeline construction authorisation (PCA) 147, 156–7
  see also water
planning contravention notices 111
planning control 86–117
  abandonment and 103
  appeals 108–11
    judicial review 109–10
    procedure 108–9
    third party rights 110–11

  assumptions on compulsory purchase 127
  conservation areas 95, 106–107
  development plan 114–17
  enforcement 105, 111–14
  General Permitted Development Order (GPDO) 88–96
    agricultural buildings and operations 92–4
    Article 4 Directions 95
    caravans 91–2
    changes of use 90–91
    forestry buildings and operations 94
    minor works 90
    nature of order 88
    other authorised development 95–6
    other restrictions 95
    single private dwellinghouse 88–90
    temporary buildings and uses 91
  government guidance 115, 117
  hedgerows 223
  listed buildings 90, 95, 103, 104–106
    listed building consent 105
  other designated property 107–108
  planning permission
    application 96–7
    conditions 97, 100–102
    fees 97
    horse grazing 102–103
    landlord's consent for 24–5
    need for 87–8
    obligations 97–9
    subject to agricultural occupancy condition 100–102
    time limits 99–100
  Tree Preservation Orders 106
planning obligations 97–9
ploughing, rights of way and 74–7
poaching
  fish 199
  game 189–91
  penalties 205–206
Pointe Gourde principle 127–8
policy planning guidance notes (PPGs) 114, 117
pollution *see* environmental protection
positive covenant 5

Index

possession, grounds for 51–2, 55–7
possessory title 9
private nuisance 163–8
  definition 163
  main rules 163–5
  remedies 167–8
  types of 165–7
proprietorship register 9
protected shorthold tenancies 58
public access 69–85
  open land and registered common land 69, 80–84
  rights of way 4, 69–72
    limitations on land use caused by 72–80
  town and village greens 69, 84–5
public fishing 195
public nuisance 162
public path orders 71
public works, compensation for use of 133–4

qualified title 9
quarries 79

rabbit clearance order 165
rating 254–61
raves 172
recreational clubs, rating and 256–7
rectification of title 10
redundancy 247–51
  consultation with trade unions 250
  death of employer and 249
  employees leaving before notice expires 249
  insolvent employer 251
  notice to Secretary of State 250
  payments 247–9, 251
  time limit for claims 251
  time off to look for work 249
  unfair selection for 249–50
re-entry, rights of 3
regional planning guidance (RPG) 114
registered common land 69, 80–84
registration of title 6–11
  business tenancies 68
  compulsory 6
  indemnity and 10
  minor interests 7–8
  open register 6

overriding interests 7
parts of register 8–9
rectification of 10
registered interests 6–7
remediation notices 214–15
remedies
  poaching 199
  private nuisance 167–8
  unfair dismissal 246–7
rent
  agricultural holdings 34–5, 44–5
    notice to pay 38
  arbitration 21, 22–3, 34, 35, 44–5
  business tenancies 64–5
  dispute resolution 27–8
  farm business tenancies 19–23
  milk quotas and 231
  residential protection of farm workers and 49–50, 55, 58
  review of *see* review of rent
Rent Assessment Committee 58
rent charges 3
residential protection of farm workers 47–59
  Housing Act 1988 53–6
    application 53
    assured agricultural occupancy 53–4
    cessation of agricultural worker condition 54–5
    continuing protection 54
    grounds for possession 55–6
    rehousing provisions 56
    rent control 55
    security of tenure 55
    succession rights 54
  Protection from Eviction Act 1977 57
  Rent Act 1977 56–7
  Rent (Agriculture) Act 1976 47–53
    application 47
    continuation of protection 49
    grounds for possession 51–2
    nature of statutory tenancy 49
    other terms 50
    qualifying ownership 48–9
    qualifying worker 47–8
    relevant licence or tenancy 49
    rent 49–50
    security of tenure 50–51

succession rights 52–3
shorthold tenancies 57–9
  purpose 57–8
  rent control 58
  security of tenure 59
  types of 58
restricted byways 70
restrictions, minor interests and 8
restrictive covenants 5–6, 7
retirement, succession of tenancies on 45
reversion
  option to purchase 7
  severed *see* severed reversion
review of rent
  agricultural holdings 34
  farm business tenancies 20, 22
rights of way 4, 69–72
  dedication and acceptance 71–2
  definitions 69
  fly tipping 216–17
  limitations on land use caused by 72–80
  obstruction 72–3
  ownership 69
  public path orders 71
  recording 69–71
  structures across 73–4
rivers 181–2
  ownership of fishing rights 193–5
  riparian rights 175–6
Royal Institution of Chartered Surveyors (RICS) 22, 25, 28, 29, 35
Rural Payments Agency (RPA) 224

safety at work 237, 253
security of tenure
  agricultural holdings 36–9
  residential protection of farm workers and 50–51, 55, 59
seizure, powers of 191, 203–204
self-employed status 236–8
self-help remedies 167–8
service of notices 29–30
severance, compulsory purchase and 130
severed reversion
  compensation for improvements and 27
  rent and 20–21
  termination of tenancies and 18
sewage sludge 211, 221
sewerage 139, 152, 153–6
share farming agreements, milk quotas and 229
shooting 187
  rights of way and 78
shorthold tenancies 57–9
  purpose 57–8
  rent control 58
  security of tenure 59
  types of 58
shotgun licence 191–2
sick pay scheme 240
silage 208
sister, succession of tenancies to 42
sites of special scientific interest (SSSI) 95, 107, 108
sludge from sewage 211, 221
slurry 208
smells 165
snares 193
soil protection 211
sport 185–206
  firearms 78, 187, 191–3
  hunting 200–201
  protection of wildlife 187–8
  sporting rights 185–7
    rating and 259
  see also fishing; game
spot listing 104
spouse, succession of tenancies to 42, 52
spray irrigation 180
stables 103
stamp duty, business tenancies 67–8
statutory nuisance 162–3
stiles 73–4
stocking of fishing lakes 181
straying animals 169–71
strict settlement 7
strict settlements 3
structure plan 114, 115
subtenancies
  agricultural holdings 32
  milk quotas and 229
succession tenancies
  agricultural holdings 42–5
  residential protection of farm workers and 52–3, 54

survey, power to 143
syndicates 258

taxation 236
  council tax 261–4
  rating 254–61
telecoms 138–9, 143, 144–6
temporary buildings and uses, planning control 91
tenants and tenancies 1
  joint tenants 1
  leasehold 2–3
  legal nature of tenancy 61–2
  tenants in common 1
  see also agricultural holdings; business tenancies; farm business tenancies; subtenancies; termination of tenancies
termination of employment
  dismissal 244–7, 252
  notice of 240, 243–4
  redundancy 247–51
termination of tenancies 17–19, 62
  break clauses 18
  business tenancies 63
  fixed term tenancies 17, 61
  lease for lives 18
  notice
    agricultural holdings 36–8
    farm business tenancies 17–19
    periodic tenancies 17–18, 61
    severed reversion 18
theft 190
third parties
  appeal rights in planning control 110–111
  dispute resolution and 28–9
time limits
  planning permission 99–100
  redundancy claims 251
time off to look for work 249
tipping *see* fly tipping
title
  registration *see* registration of title
  types of 9–10
town greens 69, 84–5
trade unions, consultation with 250
traps 193
treat, notice to 122

trees
  lopping for utility companies 144
  nuisance caused by 166
  rights of way and 77–8
  Tree Preservation Orders 106
trespass 161, 171–3, 190, 191, 200
trusts 2, 3
  family equitable interests under 4–5
  minor interests 7

unfair dismissal 244–5, 252
  remedies 246–7
unfair selection for redundancy 249–50
unoccupied property, rating and 256
utilities 137–57
  easements for 4, 137, 142
  electricity 139, 140–144
  gas 139, 146–9
  other pipelines 139, 156–7
  sewerage 139, 152, 153–6
  telecoms 138–9, 143, 144–6
  water resources and drainage 139, 149–52
  water supply 139, 152–3, 154–6
  wayleaves 137–8, 141–3, 144–5
  works in highway land and private tracks 138–40, 149

vehicles, abandoned 217
vicarious liability 237
village greens 69, 84–5
visitors
  duty owed to 159–60
    agreements changing the duty 160–161

wages 243
wander, right to 69
waste disposal 218–23
  definitions 218–20
  duty of care 222–3
  exemptions 220–221
  waste management licence 221–2
  see also fly tipping
waste plans 116
water
  abstraction of 176–80
  drainage 139, 149–52
    flood defence and 181–4
  impounding 180–181

land acquisition 150–151, 154–5, 175
pollution 207–209
  offences 207–208
  preventative measures 208, 211
  protection areas 209
responsibility for water management 174–5
riparian rights 175–6
sewerage 139, 152, 153–6
stocking fishing lakes 181
water resources 139, 149–52

water supply 139, 152–3, 154–6
wayleaves 137–8, 141–3, 144–5
weeds, nuisance caused by 166–7
wildlife
  damage caused by 41, 165–6, 185–6
  protection of 187–8
William I, King 1
wires across right of way 73–4
work notice procedures 216
World Heritage sites 108
written statement of employment 239–41